T0332933

EVOLUTION AND CONTROL IN BIOLOGICAL SYSTEMS

THE INTERNATIONAL INSTITUTE FOR APPLIED SYSTEMS ANALYSIS

is a nongovernmental research institution, bringing together scientists from around the world to work on problems of common concern. Situated in Laxenburg, Austria, IIASA was founded in October 1972 by the academies of science and equivalent organizations of twelve countries. Its founders gave IIASA a unique position outside national, disciplinary, and institutional boundaries so that it might take the broadest possible view in pursuing its objectives:

To promote international cooperation in solving problems from social, economic, technological, and environmental change

To create a network of institutions in the national member organization countries and elsewhere for joint scientific research

To develop and formalize systems analysis and the sciences contributing to it, and promote the use of analytical techniques needed to evaluate and address complex problems

To inform policy advisors and decision makers about the potential application of the Institute's work to such problems

The Institute now has national member organizations in the following countries:

Austria
The Austrian Academy of Sciences

Bulgaria
The National Committee for Applied Systems Analysis and Management

Canada
The Canadian Committee for IIASA

Czechoslovakia
The Committee for IIASA of the Czechoslovak Socialist Republic

Finland
The Finnish Committee for IIASA

France
The French Association for the Development of Systems Analysis

German Democratic Republic
The Academy of Sciences of the German Democratic Republic

Federal Republic of Germany
Association for the Advancement of IIASA

Hungary
The Hungarian Committee for Applied Systems Analysis

Italy
The National Research Council

Japan
The Japan Committee for IIASA

Netherlands
The Foundation IIASA–Netherlands

Poland
The Polish Academy of Sciences

Sweden
The Swedish Council for Planning and Coordination of Research

Union of Soviet Socialist Republics
The Academy of Sciences of the Union of Soviet Socialist Republics

United States of America
The American Academy of Arts and Sciences

EVOLUTION AND CONTROL IN BIOLOGICAL SYSTEMS

Proceedings of the IIASA Workshop, Laxenburg, Austria, 30 November – 4 December 1987

Edited by

A. B. KURZHANSKI and K. SIGMUND
International Institute for Applied Systems Analysis (IIASA),
Laxenburg, Austria

Reprinted from *Acta Applicandae Mathematicae*, Vol. 14, Nos. 1 & 2 (1989)

KLUWER ACADEMIC PUBLISHERS

DORDRECHT / BOSTON / LONDON

INTERNATIONAL INSTITUTE FOR APPLIED SYSTEMS ANALYSIS

Library of Congress Cataloging in Publication Data

IIASA Workshop (1987 : Laxenburg, Austria)
    Evolution and control in biological systems : proceedings of the
IIASA Workshop, 30 November-4 December 1987, Laxenburg, Austria /
edited by Alexander B. Kurzhanski and Karl Sigmund.
        p.   cm.
    "Reprinted from Acta applicandae mathematicae, vol. 14, nos. 1-?
(1989)"
    ISBN 0-7923-0219-2
    1. Biological control systems--Congresses. 2. Evolution-
-Congresses.   I. Kurzhanskiĭ, A. B. II. Sigmund, Karl, 1945-   .
III. Title.
QH508.I37 1987
574'.01'51--dc19   89-2401

Published by Kluwer Academic Publishers,
P.O. Box 17, 3300 AA Dordrecht, The Netherlands.

Kluwer Academic Publishers incorporates the publishing programmes of
D. Reidel, Martinus Nijhoff, Dr W. Junk and MTP Press.

Sold and distributed in the U.S.A. and Canada
by Kluwer Academic Publishers,
101 Philip Drive, Norwell, MA 02061, U.S.A.

In all other countries, sold and distributed
by Kluwer Academic Publishers Group,
P.O. Box 322, 3300 AH Dordrecht, The Netherlands.

Table of Contents

V. SIMULATION AND CONTROL IN MEDICINE

Acta Applicandae Mathematicae **14** (1989), 1.
© 1989 *by IIASA.*

INTRODUCTION

For several years IIASA has hosted or co-sponsored international workshops dealing with dynamical systems and their applications: in 1984 "Dynamics of Macrosystems" at Laxenburg (Austria), in 1985 "The Mathematics of Dynamical Systems" in Sopron (Hungary), and in 1986 "Dynamical Systems and Environmental Models" at the Wartburg (GDR).

The latest workshop, held at Laxenburg in December 1987, dealt mostly with models from biology. This reflects the fact that while classically, the theory of dynamical systems drew most of its motivations from physics, and in particular from mechanics (celestial or otherwise), there is now a growing trend to deal with problems suggested by biological models.

The present volume contains the Proceedings of this workshop.

Part I deals with ecological models. Since the "golden age" of biomathematics in the twenties and thirties, population ecology is one of the most active branches of this science. The many concepts of stability, resilience, persistence etc., and the interesting structures displayed by interactions between predators and prey, the effects of competition, immigration etc., constitute a rich field for theoretical investigation.

Another classical field of biomathematics is population genetics. In Part II, questions of frequency dependent selection and mutation - selection models are analyzed. Part III is devoted to the complexities of demographical models of populations structured by sex, age and other characters.

A rapidly growing field of applications is offered by immunology. This is covered in Part IV. The fact that apart from some basic tenets like clonal selection and the existence of T cells and B cells, there is no universally agreed theoretical model, makes the mathematical investigation of competing theories all the more important. Part V finally, deals with practical issues of simulation and control in biomedicine.

We hope that this volume illustrates the wide range of applications and motivations provided by biological systems to mathematical modelling and the theory of dynamical systems.

A.B. Kurzhanski
Chairman
and
K. Sigmund
System and Decision Sciences Program
International Institute
for Applied Systems Analysis

Acta Applicandae Mathematicae **14** (1989), 3–10.

Stability Conditions for Two Predator One Prey Systems

M. Farkas

Department of Mathematics

Budapest University of Technology

Budapest, Hungary H-1521

H.I. Freedman

Department of Mathematics

University of Alberta

Edmonton, Canada T6G2G1

AMS Subject Classification (1980): 92A17
Key words: predator-prey systems, asymptotical stability, global stability

1. Introduction

Consider the predator-prey system

$$\dot{x} = xF(x,y) , \quad \dot{y} = yG(x,y) \tag{1.1}$$

where $F, G \in C^1$,

$$F(0,0) > 0 , \ \exists K > 0 : F(K,0) = 0, \ F_x(x,0) \leq 0, \ F_y(x,y) < 0, \tag{1.2}$$

$$G(0,y) < 0, \ G_x(x,y) > 0, \ G_y(x,y) \leq 0. \tag{1.3}$$

Assume that there exists an equilibrium $E = (x^o, y^o)$ in the interior of the positive quadrant of the x, y plane, i.e. $x^o, y^o > 0$, $F(x^o, y^o) = G(x^o, y^o) = 0$.

The famous Rosenzweig-MacArthur "graphical criterion" of stability in a somewhat generalized form states that

(i)   if $F_x(x^o, y^o) < 0$ then E is asymptotically stable,

(ii) if $G_y(x^o, y^o) < 0$ and $F_x(x^o, y^o) = 0$ then E is asymptotically stable,

(iii) if G does not depend on y and $F_x(x^o, y^o) > 0$ then E is unstable,

(iv) if G does not depend on y, and by varying a parameter in the system we get $F_x(x^o, y^o) = 0$ then at this value of the parameter E undergoes an Andronov-Hopf bifurcation.

(Cf. [4],[2]).

This criterion can be interpreted intuitively as follows. Figure 1(a) illustrates (i): if the system is moved out of the equilibrium by either increasing-decreasing the quantity of prey or increasing-decreasing the quantity of predators into the respective points A,B,C, or D then the dynamics acts to *decrease* the distance from E.

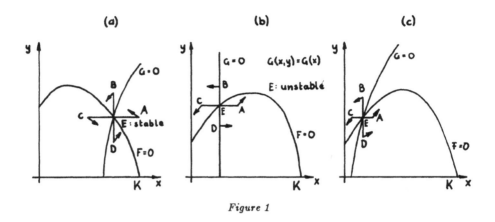

Figure 1

Figure 1(b) illustrates (iii): if G does not depend on y, and the system is moved out of the equilibrium into A,B,C or D then the dynamics acts to *increase* the distance from E. Figure 1(c) illustrates that (iii) *is not true necessarily* if G depends on y because at A and C the dynamics drives the system away from E and at Band D drives it closer to E. One could interpret case (ii) by a similar figure but it would need a little bit more imagination. We note that in the classical Lotka-Volterra case when $F(x,y)=1-y$, $G(x,y) = -1+x$ the corresponding figure shows clearly that a neutrally stable situation can be expected.

The question arises whether this simple, meaningful stability criterion can be generalized to the three dimensional case when two predators compete for a single prey species, i.e. to the system

$$\dot{x} = xf(x,y_1,y_2)$$

$$\dot{y}_1 = y_1 f_1(x,y_1,y_2) \tag{1.4}$$

$$\dot{y}_2 = y_2 f_2(x, y_1, y_2)$$

where, as in the whole paper, x denotes the quantity of prey, y_i denotes the quantity of the i-th predator and the functions f, g_i satisfy natural conditions (i=1,2).

We are going to show that the criterion cannot be generalized to the general system but can be in the important special case in which there is no interspecific competition between the predators, i.e. g_i does not depend on y_{3-i}, (i=1,2). We are also establishing a global stability criterion for the latter case. The theorems are given here without proofs. The proofs can be found in [1].

2. Generalization of the graphical stability criterion

Consider the system

$$\dot{x} = x F(x, y_1, y_2, K)$$

$$\dot{y}_1 = y_1 G_1(x, y_1) \tag{2.1}$$

$$\dot{y}_2 = y_2 G_2(x, y_2)$$

where $F \in C_1[\mathbf{R}_+^3 \times \mathbf{R}_+]$, $G_i \in C^1[\mathbf{R}_+^2]$, ($i$=1,2),

$$F(0,0,0,K) > 0, \quad F(K,0,0,K) = 0, \tag{2.2}$$

$$F_x(x,0,0,K) \le 0, \quad F_{y_i} < 0, \tag{2.3}$$

$$G_i(0,y_i) < 0, \quad G_{ix} > 0, \quad G_{iy_i} \le 0, \tag{2.4}$$

$$G_{1y_1}^2 + G_{2y_2}^2 > 0. \tag{2.5}$$

Conditions (2.2) - (2.4) express the fact that x is the quantity of prey, y_i is the quantity of the i-th predator, K is the carrying capacity of the environment with respect to the prey. (2.5) is a technical condition without which we could prove essentially the same results but in more complicated form.

Besides the conditions above, we assume that to each value of predator quantity there belongs a threshold prey quantity above which the predator quantity grows, i.e.

$$\exists \lambda_i : \mathbf{R}^+ \mapsto \mathbf{R}^+, \quad \lambda_i \in C^1 : G_i(\lambda_i(y_i), y_i) \equiv 0, \tag{2.6}$$

and that there exists an equilibrium in the positive orthant, i.e.

$$\forall K > K_o \geq 0, \ \exists E : (x^o(K), y_1^o(K), y_2^o(K)),$$

$$F(E,K) = 0, \ G_i(E) = 0, \ x^o, y_i^o \geq 0, \ (i = 1,2) . \tag{2.7}$$

Under these assumptions the following theorem can be proved.

Theorem 2.1. If the equilibrium E of system (2.1) is in the interior of the positive orthant, and

$$F_x(E,K) \leq 0 \tag{2.8}$$

then E is asymptotically stable.

A typical situation is shown on Figure 2. The set where $G_i(x,y_i) = 0$ is a cylinder in the three dimensional space based on the curve $x = \lambda_i(y_i)$, with generators parallel to the axis $y_{3-i}, (i = 1,2)$. These two cylinders intersect in a curve which in its part intersects the level surface $F(x,y_1,y_2,K) = 0$. The point of intersection is E. It is *asymptotically*

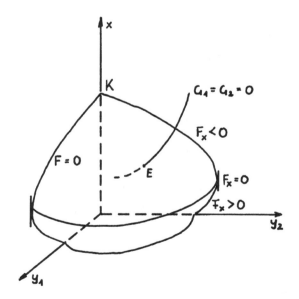

Figure 2

stable if it falls upon the "northern hemisphere" (including the "equator" of the onion-like level surface $F = 0$. It *might be unstable* if it falls upon the "southern hemisphere." When K is increased the point E may move "southwards" on the surface $F = 0$ and at a certain value of K may lose its stability. (The fact that this does not happen yet at reaching the equator is due to condition (2.5).) In the special case when there is *no intraspecific competition at one of the predators* we can characterize this bifurcation.

Consider the system

$$\dot{x} = xF(x,y_1,y_2,K)$$

$$\dot{y}_1 = y_1 G_1(x), \quad \dot{y}_2 = y_2 G_2(x,y_2) \tag{2.9}$$

under the conditions (2.2) - (2.7).

Theorem 2.2. If with increasing K the equilibrium E of (2.9) stays in the interior of the positive orthant of the (x,y_1,y_2)-space and loses its stability then this happens by an Andronov-Hopf bifurcation.

Note that if intraspecific competition is lacking for *both* predators then in the generic case there will be no equilibrium in the positive orthant.

3. Examples

Consider the system

$$\dot{x} = x(K - x - \frac{y_1 + 2y_2}{2(x + 1)})$$

$$\dot{y}_1 = y_1(-1 + \frac{9x}{8(x + 1)}) \tag{3.1}$$

$$\dot{y}_2 = y_2(-1 + \frac{4x}{x + 1} - y_2)$$

which satisfies conditions (2.2) - (2.7) for $K > 1342/162 = 8.284$. The equilibrium of this system $E(K) : (8 ; 18K - 1342/9; 23/9)$ is in the interior of the positive orthant for values of K specified above. It can be shown that E is asymptotically stable for $8.284 < K < 17.043$ and that it undergoes an Andronov-Hopf bifurcation at $K = 17.043$. Note that $F_x(E(K),K) > 0$ for $17 < K < 17.043$ showing that $F_x \leq 0$ is *not necessary* for asymptotic stability. Note also that the system seems to satisfy the conditions of persistence due to Freedman and Waltman [3] for $8.284 < K < 17$.

In order to show that the Rosenzweig-MacArthur condition is not valid for general systems of form (1.4) consider the system

$$\dot{x} = xF(x,y_1,y_2,K)$$

$$\dot{y}_1 = y_1 G_1(x,y_2), \quad \dot{y}_2 = y_2 G_2(x,y_1) \tag{3.2}$$

i.e. the system where there is interspecific competition but no intraspecific competition among the predators. Conditions (2.2) - (2.7) are assumed to hold "mutatis mutandis," e.g. the coordinates of the equilibrium $E : (x^o, y_1^o, y_2^o)$ satisfy $x^o, y_1^o, y_2^o > 0$, $F(x^o, y_1^o, y_2^o, K) = 0$, $G_1(x^o, y_2^o) = G_2(x^o, y_1^o) = 0$. One can show easily that in this case E is *always unstable* irrespective of the sign of F_x.

4. Global Stability

A fairly general and realistic specialization of system (2.1) is obtained if we assume that the functional responses occurring in the equations concerning the predators are present with some predation rate factors and with a negative sign in the equation concerning the prey. Consider the system

$$\dot{x} = xF(x,y_1,y_2), \quad \dot{y}_1 = y_1 G_1(x,y_1), \quad \dot{y}_2 = y_2 G_2(x,y_2) \tag{4.1}$$

where

$$xF(x,y_1,y_2) = xh(x) - k_1 y_1(G_1(x,y_1) - G_1(0,y_1)) - k_2 y_2(G_2(x,y_2) - G_2(0,y_2)), \quad k_i > 0.$$

We assume that besides conditions (2.3) - (2.7) we have

$$h(0) > 0; \quad h'(x) < 0, \quad x \in \mathbf{R}_+; \quad \exists K > 0, \quad h(K) = 0$$

$$G_i(0,0) < 0, \quad G_i(K,0) > 0, \quad G_{iy_i} < 0$$

and that $|G_i(0,y_i)|$ is bounded $(i = 1,2)$.

One may prove that the set

$$A = \{(x,y_1,y_2): 0 \le x \le K, 0 \le x + k_i y_i \le -\frac{M_i}{G_i(0,0)}, i = 1,2\}$$

where

$$M_i = sup(xh(x) - xG_i(0,y_i)), \quad 0 \le x \le K, \quad 0 \le y_i, \quad (i = 1,2).$$

is positively invariant and globally attractive with respect to system (4.1). Then applying the Ljapunov function

$$V(x,y_1,y_2) = \int_{x^o}^{x} \left[\frac{G_1(s,y_1^o)}{G_1(s,y_1^o) - G_1(0,y_1^o)} + \frac{G_2(s,y_2^o)}{G_2(s,y_2^o) - G_2(0,y_2^o)} \right] ds$$

$$+ k_1[y_1 - y_1^o - y_1^o \ ln \ \frac{y_1}{y_1^o}] + k_2[y_2 - y_2^o - y_2^o \ ln \ \frac{y_2}{y_2^o}]$$

it turns out that it is positive definite relative to the equilibrium $E:(x^o,y_1^o,y_2^o)$. Its derivative with respect to system (4.1) is -1 times a "pseudoquadratic form" whose coefficients are

$$a_{oo}(x) = - \ \frac{1}{(x-x^o)^2} \ [\ \frac{G_1(x,y_1^o)}{G_1(x,y_1^o) - G_1(0,y_1^o)} + \frac{G_2(x,y_2^o)}{G_2(x,y_2^o) - G_2(0,y_2^o)} \ ] \ xF(x,y_1^o,y_2^o),$$

$$a_{ii}(x,y_i) = -k_i(G_i(x,y_i) - G_i(x,y_i^o))/(y_i - y_i^o),$$

$$a_{oi}(x,y_i) = - \ \frac{k_i}{(x - x^o)(y_i - y_i^o)} \ [\ y_i G_i(x,y_i^o)(1 - \frac{G_i(x,y_i) - G_i(0,y_i)}{G_i(x,y_i^o) - G_i(0,y_i^o)})$$

$$+ \ \frac{G_{3-i}(x,y_{3-i}^o)}{G_{3-i}(x,y_{3-i}^o) - G_{3-i}(x,y_{3-i}^o)} \ (y_i^o(G_i(x,y_i^o) - G_i(0,y_i^o)) - y_i(G_i(x,y_i) - G_i(0,y_i))) \ ],$$

$$(i = 1,2) \ .$$

This way one can prove

Theorem 4.1. If for $(x,y_1,y_2) \in A$

$$a_{22}(4a_{oo}a_{11} - a_{01}^2) - a_{02}^2 a_{11} > 0 \qquad (4.2)$$

then the equilibrium $E : (x^o,y_1^o,y_2^o)$ is globally asymptotically stable with respect to Int $\mathbf{R}_+^3$.

We note that the conditions imposed upon the system imply $a_{ii} > 0$, $(i = 1,2)$. As a consequence, (4.2) implies $a_{oo} > 0$, i.e. the latter inequality is a necessary condition for (4.2) to hold.

Some relation can be seen between the global stability condition (4.2) on the one hand and the local stability condition (2.8) on the other. Namely, the "strengthened condition" (2.8) $F_x < 0$ implies $a_{oo} > 0$ *locally* in a neighborhood of x^o, and inversely, $a_{oo}(x) > 0$ implies that $xF(x,y_1^o,y_2^o)$ is locally decreasing at x^o, i.e. $F_x(x^o,y_1^o,y_2^o) \leq 0$.

5. References

[1] Farkas, M., Freedman, H.I.: The stable coexistence of competing species on a renewable resource, to appear in *J. Math. Anal. Appl.*

[2] Freedman, H.I.: Graphical stability, enrichment, and pest control by a natural enemy, *Math. Biosci.* 31(1976) 207-225.

[3] Freedman, H.I., Waltman, P.E.: Persistence in models of three interacting predator-prey populations, *Math. Biosci.* 68(1984) 213-231.

[4] Rosenzweig, M.L., MacArthur, R.H.: Graphical representation and stability conditions of predator-prey interactions, *Amer. Nat.* 47(1963) 209-223.

Acta Applicandae Mathematicae **14** (1989), 11–22.
© 1989 *by IIASA.*

A Unified Approach to Persistence

Josef Hofbauer

Institut für Mathematik
Universität Wien
Strudlhofgasse 4
A-1090 Wien, Austria

AMS Subject Classification (1980): 92A17, 34D20
Key words: repeller, persistence, Morse decomposition, time averages, permanence

1. Repellers in dynamical systems

Let f^t be a flow on a compact metric space X and M be a closed invariant subset of X.

Theorem 1. *If M is isolated then one of the following three alternatives hold.*

(a) *M is an attractor.*

(b) *M is a repeller.*

(c) *M is a 'saddle': there exist $x, y \notin M$ such that $\omega(x) \subset M$ and $\alpha(y) \subset M$.*

Remark. 'Attractor' stands here for the 'stable attractor' of [1, ch.V], or as used by Conley [5]. In case (b), when M is a repeller, there is a dual attractor which attracts all orbits in $X \setminus M$ (see [5, ch.II.5]).

Actually, we need only the following special case.

Theorem 2. *M is a repeller if and only if*

(1) *M is isolated, and*

(2) *$W^s(M) \subset M$, so that no orbit from $X \setminus M$ converges to M.*

Remark. For completeness we sketch the (simple) proof. Choose a compact isolating neighborhood U of M. $\omega(x)$ of any $x \notin M$ must meet $X \setminus U$ if neither (1) nor (2) is violated. Then $X \setminus U$ is a weakly attracting region for $X \setminus M$. Its forward invariant closure $\gamma^+(\overline{X \setminus U})$ is then still compact and contains an attractor for $X \setminus M$ whose dual repeller is M. [Consult the 'weak attractor theorem' of [1, ch. V, 1.25] or, in a more modern and concise language, Conley [5, II.5.1.D] (with reversed time) or Hutson's [15] Lemma 2.1 for details]. The general result, Theorem 1, follows immediately from Theorem 2 and its time reversal.

COROLLARY 1. *Let* $P : X \to \mathbb{R}$ *be a continuous function on* X *satisfying the following conditions.*

(a) $P(x) = 0$ *for* $x \in M$ *and* $P(x) > 0$ *for* $x \in X \setminus M$.

(b) *For every* x *in a neighbourhood* U *of* M *there exists a time* $t > 0$ *such that* $P(xt) > P(x)$. *Then* M *is a repeller.*

Proof: Suppose, for some $x \in U \setminus M$, $\gamma^+(x) \subset U$. Then there is a $y \in \overline{\gamma^+(x)}$ such that $P(y) \geq P(z)$ for all $z \in \gamma^+(x)$ and in particular $P(y) \geq P(yt)$ for $t \geq 0$. This contradicts (b). Hence U is an isolating neighbourhood of M and for no $x \in U \setminus M$, $\omega(x) \subset U$. Hence, by Theorem 2, M is a repeller. $\square$

As shown by Fonda [6] this condition readily implies those given earlier by Gard and Hallam [9] in terms of 'persistence functions' and Hofbauer [11] and Hutson [15] in terms of 'average Ljapunov functions' which are more accessible to concrete applications, since they involve only conditions for $x \in M$:

COROLLARY 2. *If* P *is differentiable along orbits then condition (b) can be replaced by*

(b') *There is a continuous function* $\psi : X \to \mathbb{R}$, *such that* $\dot{P}(x) \geq P(x)\psi(x)$ *for all* $x \in X$, *and for each* $x \in M$ *there is a time* $T > 0$ *such that*

$$\int_0^T \psi(x(t))\, dt > 0. \tag{1}$$

It is also sufficient that (1) holds for all $x \in \omega(M)$.

Up to now no information on the flow on M was assumed or required. However, sometimes it is useful to exploit the structure of the flow on M. The main idea is, if an orbit $x \in X \setminus M$ converges to M then its ω-limit cannot be any closed invariant subset of M, but has the crucial property of being *chain transitive* (see [2]). To explain this we need some notation.

An $(\varepsilon, T)-pseudoorbit$ or $(\varepsilon, T)-chain$ for f^t is the union $\bigcup_{i=1}^n x_i[0, \tau_i]$ of pieces of orbits of length $\tau_i \geq T$ and the jumps obeying $d(x_i\tau_i, x_{i+1}) < \varepsilon$. If the flow f^t is given by a C^1 vector field $\dot{x} = g(x)$ then one could use $\varepsilon-approximate$ *solutions* $\{y(t) : 0 \leq t \leq T'\}$ satisfying $|\dot{y}(t) - g(y(t))| < \varepsilon$ instead. We say x *is chained to* y if for all (small) $\varepsilon > 0$ and all (large) $T > 0$ there is an (ε, T)-chain with $x_0 = x$ and $x_{n+1} = y$. If x is chained to x, then it is said to be a *chain recurrent* point, and we write $x \in \mathcal{R}$. Now $x \notin \mathcal{R}$ iff there exists an open set U such that $f^t \bar{U} \subset U$ $\forall t > 0$ and $x \in U \setminus f^t U$ for some $t > 0$, or equivalently, iff there exists an attractor A, such that $x \in W^s(A) \setminus A$. Thus $\mathcal{R} = \bigcap A \cup A^*$, the intersection over all attractor-repeller pairs (A, A^*). The connected components Λ_i of $\mathcal{R}$, sometimes called the *basic sets* of the flow, are the maximal *chain transitive* subsets of X: for all $x, y \in \Lambda_i$, x is chained to y, with chains lying completely in Λ_i.

This is the essence of the *structure theorem* of Conley [5] for general dynamical systems. In the following theorem we assume that the structure of the flow f^t on M is given.

Theorem 3. *Let Λ_i be the basic sets of M. Then M is a repeller iff each basic set Λ_i of M satisfies the following two conditions.*

(1) *Λ_i is M-isolated in X, in the sense that some neighbourhood U of Λ_i in X contains a full orbit $\gamma(x)$ only if $x \in M$.*

(2) *$W^s(\Lambda_i) \subset M$.*

In some sense this result is best possible since a priori each basic set in M is a candidate for an ω-limit set of an orbit $x \in X \setminus M$.

Proof. That (1) and (2) are necessary conditions, is clear. For the converse we have to show that (1) and (2) of Theorem 2 are fulfilled. As already pointed out, (2) is a consequence of the chain transitivity of ω-limit sets. It remains to show that M is isolated. Suppose not. Then there exists a sequence of compact invariant sets $\Gamma_k \not\subset M$ which can be chosen to be chain transitive such that $\Gamma_k \subset \{x : d(x, M) \leq \frac{1}{k}\}$. A subsequence of (Γ_k) will converge in the Hausdorff metric to a compact set $\Gamma \subset M$ which is again invariant and chain transitive. (See the appendix in [18] for details.) Thus Γ must be contained in a single basic set Λ_i. But then any neighborhood of Λ_i contains some Γ_k for large k. Hence Λ_i is not M-isolated. $\square$

Basic sets need not be isolated in M (not even generically) and there may even be uncountably many of them. Hence it is often difficult to determine the complete decomposition into basic sets. Usually it is more convenient to find a finite covering $\{M_1, M_2, ..., M_n\}$ such that each basic set lies in one M_i. Such a *Morse decomposition* into closed invariant sets $\{M_1, \ldots, M_n\}$ is characterized by the property that for each $x \in M$, either $x \in M_i$ for some i, or $\omega(x) \subset M_i$ and $\alpha(x) \subset M_j$ for some $i < j$.

COROLLARY 3. *Let $\{M_1, \ldots, M_n\}$ be a Morse decomposition for M such that each Morse set M_i is also isolated in X. Then M is a repeller if and only if no M_i attracts orbits from $X \setminus M$, i.e. for each M_i, $W^s(M_i) \subset M$.*

Remark. This is essentially the Butler-Freedman-Waltman persistence theorem (see Theorem 3.1 of [4]). Their acyclic covering of the ω-limit sets of M is indeed a Morse decomposition, as shown e.g. in [7]. The proof of this uses the so-called Butler-McGehee lemma, which by itself is a simple consequence of the

Ura-Kimura theorem. We conclude with a simple combination of Corollary 1
and Theorem 3.

COROLLARY 4. *Suppose that for each basic set Λ (or each Morse set M_i of a
Morse decomposition of M) there is a function P which is defined in a neighbourhood U of Λ (resp. M_i) and satisfies the conditions of Corollary 1 or 2 in
this neighbourhood U. Then M is a repeller.*

In applications to ecological problems one usually encounters the nonnegative orthant $\mathbb{R}^n_+$ as state space, and a dissipative flow thereon. Considering
the one-point compactification $X = \mathbb{R}^n_+ \cup \{\infty\}$, the system is *permament* or
uniformly persistent iff the closed invariant subset $M = \mathrm{bd}\mathbb{R}^n_+ \cup \{\infty\}$ is a repeller. Another way would be to take X_1 as the set of all bounded orbits in
$\mathbb{R}^n_+$, that is the dual attractor of the repeller $\{\infty\}$, and M_1 the dual attractor
to $\{\infty\}$ in $\mathrm{bd}\mathbb{R}^n_+$.

Corollary 4 above provides a natural combination of the two dual approaches to persistence problems, opened by Freedman and Waltman [8] and
Schuster and Sigmund [19], respectively. It seems the most effective for concrete
applications: Determine first the basic sets of M, or find at least a suitably fine
Morse decomposition. Then use Ljapunov type arguments to show that no orbit
$x \in X \setminus M$ converges to or even stays near one of these basic sets. For a typical
application see section 3.

Theoretically, the problem of persistence is thus reduced to the study of
the local behaviour around basic sets. In ecological applications such basic sets
- additionally to the 'classical' ones (fixed points, periodic orbits, suspensions
of subshifts, etc.) - frequently occur as 'heteroclinic cycles' (a preliminary definition might be C^1-robust chain transitive set which is not contained in the
interior of a single face of $\mathrm{bd}\mathbb{R}^n_+$). In simple cases the stability and bifurcation
of such heteroclinic cycles in arbitrary dimensions has been discussed in [12],
but further analysis is necessary.

2. Time averages and permanence of Lotka-Volterra equations

In this section we consider the familiar Lotka-Volterra equations

$$\dot{x}_i = x_i\left(r_i + \sum_{j=1}^{n} a_{ij}x_j\right) \qquad i = 1,\ldots,n \qquad (2)$$

and assume throughout that they are dissipative, i.e. their orbits are uniformly
bounded for $t \to +\infty$: there is a constant K such that

$$\limsup_{t \to +\infty} x_i(t) \leq K \qquad \text{for all } x \in \mathbb{R}^n_+.$$

Explicit criteria for this are given in [14, ch. 21.2].

There are two sufficient conditions for permanence for such Lotka-Volterra systems, see Jansen [16] and Hofbauer and Sigmund [13,14].

Theorem 4. *If the system of linear inequalities*

$$\sum_{i=1}^{n} p_i\big(r_i + (Ax)_i\big) > 0, \tag{3}$$

where x runs through all boundary fixed points of (2), admits a positive solution $p_i > 0$ then (2) is permanent.

Theorem 5. *If the convex hull C of all boundary fixed points is disjoint from the set $D = \{x \in \mathbb{R}^n_+ : r + Ax \le 0\}$ then (2) is permanent.*

The proof of Theorem 4 consists in checking that $P(x) = \prod x_i^{p_i}$ is an average Ljapunov function and satisfies (b') of Corollary 2. Theorem 5 can be reduced to Theorem 4. It is the aim of this section to give a new, independent proof of this 'geometric condition' for permanence, without recourse to average Ljapunov functions.

The key is to study the asymptotic behaviour of time averages for Lotka-Volterra equations for which we introduce the following notation:

$$m_T(x) = \frac{1}{T} \int_0^T x(t)\,dt$$

$$\mu(x) = \big\{ y : y = \lim_{T_n \to +\infty} m_{T_n}(x) \big\} \tag{4}$$

$$\mu(A) = \bigcup_{x \in A} \mu(x)$$

A classical result (going back to Volterra) states that for a 'persistent' orbit $x(t)$, $\mu(x)$ consists only of interior fixed points (in general a unique one). This follows from the identity

$$\frac{\log x_i(T) - \log x_i(0)}{T} = r_i + \big(Am_T(x)\big)_i. \tag{5}$$

If $\liminf_{t \to +\infty} x_i(T) > 0$, the left hand side goes to 0. Hence any $z \in \mu(x)$ is a fixed point of (2).

For arbitrary x this argument yields only

$$\mu(x) \subset D. \tag{6}$$

If one ignores the trivial case $\omega(x) = \{0\}$, then $0 \notin \omega(x)$ and hence $\mu(x) \subset \mathrm{bd}D$.

To get further information on $\mu(x)$ we first need a general fact on the limit sets $\mu(x)$ of time averages, valid for dynamical systems on a convex subset of a normed space.

LEMMA.
$$\mu(x) \subset \operatorname{conv} \mu(\omega(x))$$

Proof. We show that for any given $\varepsilon > 0$, $\mu(x)$ is contained in the ε-neighbourhood of $\operatorname{conv} \mu(\omega(x))$. Obviously, for each z there is a time $T = T(\varepsilon, z)$ such that $d(m_T(z), \mu(z)) < \frac{\varepsilon}{2}$. Although T is in general not a bounded function of z (e.g. near an unstable fixed point), we still can conclude the existence of a constant $\tau = \tau(\varepsilon, K)$ such that

$$\forall z \in K \quad \exists T \in [1, \tau(\varepsilon, K)] : d(m_T(z), \mu(K)) < \frac{\varepsilon}{2}$$

for any compact set K which in our case will be $\omega(x)$ or $\overline{\gamma^+(x)}$. Now choose $\delta > 0$ such that

$$|y - z| < \delta \implies |y(t) - z(t)| < \frac{\varepsilon}{2} \text{ for } 0 \le t \le \tau(\varepsilon) \text{ and } y, z \in \overline{\gamma^+(x)}$$

and hence

$$|m_T(y) - m_T(x)| < \frac{\varepsilon}{2} \text{ for } 0 \le T \le \tau(\varepsilon).$$

Now consider the orbit $x(t)$, and find a $t_0 > 0$ such that $d(x(t), \omega(x)) < \delta$ for all $t \ge t_0$. Then there exists a $z_0 \in \omega(x)$ such that $|x(t_0) - z_0| < \delta$ and hence a time $T_1 \in [1, \tau(\varepsilon)]$ with

$$|m_{T_1}(x(t_0)) - m_{T_1}(z_0)| < \frac{\varepsilon}{2} \text{ and } |m_{T_1}(z_0) - \mu(\omega(x))| < \frac{\varepsilon}{2}.$$

Thus we have a $T_1 \in [1, \tau(\varepsilon)]$ such that

$$d(m_{T_1}(x(t_0)), \mu(\omega(x))) < \varepsilon.$$

Repeating this argument, we find a point $z_1 \in \omega(x)$ near $x(t_0 + T_1)$, etc. Hence $m_T(x)$ can be represented as a convex linear combination of vectors $m_{t_0}(x), m_{T_1}(x(t_0)), m_{T_2}(x(t_0 + T_1)), \ldots$ each of which (up to the first and the last one, whose weight goes to 0 as $T \to \infty$) is ε-close to $\mu(\omega(x))$. So $m_T(x)$ itself is ε-close to $\operatorname{conv} \mu(\omega(x))$ for large T. □

Now let $\mathcal{F}_x$ denote the set of all open faces of $\mathbb{R}_+^n$ met by $\omega(x)$:

$$\mathcal{F}_x = \{I \subset \{1, \ldots, n\} : \exists z \in \omega(x), \operatorname{supp} z = I\},$$

and let $\mathcal{G}_x = \{F_I : I \in \mathcal{F}_x\}$ consist of all rest points $z = F_I$ with $r_i + (Az)_i = 0$ and $z_i \ge 0$ for $i \in I$ and $z_i = 0$ for $i \notin I$ in those faces, and let

$$C_x = \operatorname{conv} \mathcal{G}_x$$

be the convex hull of those fixed points. Then we have

Theorem 6. *For Lotka-Volterra equations:* $\mu(x) \subset C_x$.

Proof. By induction on the size of the support of x. Suppose w.l.o.g. that the statement is true for all $y \in \text{bd}\mathbb{R}^n_+$, and let $x \in \text{int}\mathbb{R}^n_+$. If $\omega(x) \subset \text{bd}\mathbb{R}^n_+$ then by induction hypothesis $\mu(y) \subset C_y \subset C_x$ for all $y \in \omega(x)$, since $\omega(y) \subset \omega(x)$. Hence $\text{conv}\,\mu(\omega(x)) \subset C_x$, and together with the lemma, $\mu(x) \subset C_x$.
If $\omega(x) \not\subset \text{bd}\mathbb{R}^n_+$ then we have to recall (5): if $\log x_i(T)/T > -\delta$ or $x_i(T) > e^{-\delta T}$ for all i then

$$|r + Am_T(x)| < \delta \qquad \text{for large } T. \tag{7}$$

For any given ε there is then a δ such that (7) implies that $m_T(x)$ is ε-close to an interior fixed point of (2). For a large time T consider now the last instance $T' \leq T$ when $x_i(T') \geq e^{-\delta T'}$ holds for all i. For the remaining time $T' < t \leq T$ at least one $x_i(t) < e^{-\delta t}$, so $x(t)$ is very close to $\text{bd}\mathbb{R}^n_+$. For this time interval we repeat the construction from the proof of the Lemma to find $m_{T-T'}(x(T'))$ ε-close to $\text{conv}\,\mu(\omega(x) \cap \text{bd}\mathbb{R}^n_+)$. Joining the two pieces puts $m_T(x)$ ε-close to C_x. $\square$

As a consequence, $\mu(x) \subset C_x \cap D$. Now if $\omega(x) \subset \text{bd}\mathbb{R}^n_+$ then $C_x \subset C$, the convex hull of *all* boundary fixed points, and hence $\mu(x) \subset C \cap D$. This shows that *no interior orbit can converge to the boundary if* $C \cap D = \emptyset$, i.e. the second part (2) of Theorem 2. The isolatedness might be shown by a similar argument. We omit this here since we hope to replace it one day by a simple robustness argument.

Hence altogether the proof of Theorem 5 presented here won't be shorter than the original one. But the result on time averages might be of interest per se. A good example to illustrate it is the May-Leonard case where $\mu(x) = C_x \cap \text{bd}D$ equals the three sides of a triangle in $\text{int}\mathbb{R}^3_+$, see [14, ch. 9.5].

It has been shown in [14, ch. 22] that the conditions of Theorems 4 and 5 characterize robust permanence for $n = 3$. As shown in the next section, this is no longer true for $n \geq 4$. Like in section 1, the condition can be weakened, however. For a basic set Λ of $\text{bd}\mathbb{R}^n_+$, let $C_\Lambda = \bigcup_{x \in \Lambda} C_x$. Then the above proof yields

Theorem 7. *Suppose for each basic set Λ of the boundary $\text{bd}\mathbb{R}^n_+$, $C_\Lambda \cap D = \emptyset$, or equivalently that (3) holds for all fixed points $x \in C_\Lambda$. Then (2) is permanent.*

There is some evidence that Theorem 7 might even be a characterization of robust permanence for Lotka-Volterra equations.

3. An example

Let us consider now the following two-prey two-predator system

$$\dot{x}_1 = x_1 \left(1 - \frac{2}{3}x_1 - \frac{1}{3}x_2 - by_1 - y_2 \right)$$

$$\dot{x}_2 = x_2 \left(1 - \frac{1}{3}x_1 - \frac{2}{3}x_2 - y_1 - by_2 \right) \tag{8}$$

$$\dot{y}_1 = y_1(-1 + 2x_1 + x_2)$$

$$\dot{y}_2 = y_2(-1 + x_1 + 2x_2)$$

This example nicely illustrates the method for proving permanence described in section 1. Furthermore it shows that the conditions given in Theorems 4 and 5 do *not characterize* permanence of Lotka-Volterra equations if $n \geq 4$. I realized this example while reading between the lines of Kirlinger's thesis [17].

(8) has the following fixed points:

$$F_1 = \left(\frac{3}{2}, 0, 0, 0 \right), \ F_{12} = (1, 1, 0, 0), \ F_1^1 = \left(\frac{1}{2}, 0, \frac{2}{3b}, 0 \right), \ F_1^2 = \left(1, 0, 0, \frac{1}{3} \right),$$

$$F_{12}^{12} = \left(\frac{1}{3}, \frac{1}{3}, \frac{2}{3(b+1)}, \frac{2}{3(b+1)} \right)$$

and their symmetric images F_2, F_2^1, F_2^2. From the external eigenvalues

$$\frac{\dot{x}_2}{x_2}(F_1^1) = \frac{5b-4}{6b}, \quad \frac{\dot{x}_1}{x_1}(F_2^1) = \frac{2-b}{3}, \quad \frac{\dot{y}_1}{y_1}(F_{12}) = 2,$$

we see that, for $b > 2$, F_2^1 is the only saturated fixed point in the subsystem $\{y_2 = 0\}$ and hence globally stable there (see Fig 1a). Since

$$\frac{\dot{y}_1}{y_1}(F_2^2) = -\frac{1}{2} < 0, \quad \frac{\dot{y}_2}{y_2}(F_2^1) = 1 > 0,$$

F_2^2 attracts all orbits in $\{x_1 = 0\}$ (Fig 1b).

In particular this shows the existence of a heteroclinic cycle γ: $F_1^1 \rightarrow F_2^1 \rightarrow F_2^2 \rightarrow F_1^2 \rightarrow F_1^1$. Since all orbits on the boundary of $\mathbb{R}_+^4$ converge to fixed points we obtain the following graphic description of the boundary flow (Fig 2), or after identifying the irreducible component corresponding to γ (Fig 3). It shows the topological structure of the boundary flow, i.e. the basic sets O, F_1, F_2, F_{12}, and γ, and the connections between them.

Since γ is a "planar" heteroclinic cycle (at each fixed point there is just one positive and one negative external eigenvalue), its stability is easily determined

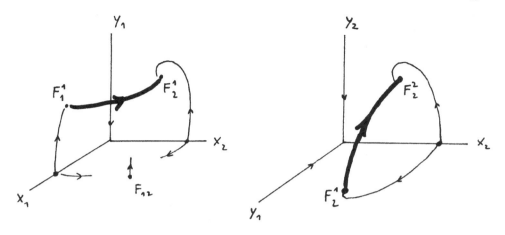

Figure 1a   Figure 1b

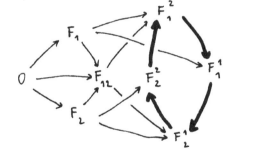

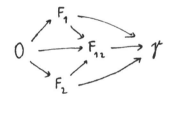

Figure 2   Figure 3

by the product of the ratios of the outgoing versus the incoming eigenvalues (see e.g. [11] or [14, ch. 22]). In view of the symmetry we only need to look at two fixed points, say F_1^1 and F_2^1):

$$\rho = \frac{5b-4}{6b} \cdot 2 \cdot \frac{3}{b-2} = \frac{5b-4}{b(b-2)}.$$

γ is attracting if $\rho < 1$, i.e. if $b > b_1 = \frac{7+\sqrt{33}}{2} \approx 6.4$, and repelling for $b < b_1$.

Hence the *system (8) is permanent for $2 < b < b_1$*, as a consequence of Theorem 3 since neither of the basic sets attracts interior orbits.

An alternative way to prove permanence of (8) would be to construct an average Ljapunov function P for the whole system and apply Corollary 2 (this way we need not know the complete structure of the basic sets, but use only the fact that all boundary orbits converge to fixed points):

$$P = (x_1 + x_2)^{p_0} x_1^{p_1} x_2^{p_2} y_1^{q_1} y_2^{q_2}$$

(this choice is motivated by Hutson [15]). Then

$$\frac{\dot{P}}{P} = p_0 \frac{\dot{x}_1 + \dot{x}_2}{x_1 + x_2} + p_1 \frac{\dot{x}_1}{x_1} + p_2 \frac{\dot{x}_2}{x_2} + q_1 \frac{\dot{y}_1}{y_1} + q_2 \frac{\dot{y}_2}{y_2}$$

Using the estimate

$$\frac{\dot{x}_1 + \dot{x}_2}{x_1 + x_2} \geq \min\left(\frac{\dot{x}_1}{x_1}, \frac{\dot{x}_2}{x_2}\right)$$

near the origin (the only point where $\dot{P}/P$ is not continuous), we need a solution of the following set of inequalities. (Those for F_1, F_2, and F_{12} are trivially satisfied.)

$$
\begin{aligned}
(O) \qquad & p_0 + p_1 + p_2 - q_1 - q_2 > 0 \\
(F_1^1) \qquad & \frac{5b-4}{6b} p_2 - \frac{1}{2} q_2 > 0 \\
(F_2^1) \qquad & -\frac{b-2}{3} p_1 + q_2 > 0 \\
(F_2^2) \qquad & \frac{5b-4}{6b} p_1 - \frac{1}{2} q_1 > 0 \\
(F_1^2) \qquad & -\frac{b-2}{3} p_2 + q_1 > 0
\end{aligned}
\qquad (9)
$$

Because of the symmetry, we can set w.l.o.g. $p_1 = p_2 = p$ and $q_1 = q_2 = q$. Then (9) leads to

$$\frac{5b-4}{6b} > \frac{q}{p} > \frac{b-2}{3} \qquad \text{and} \qquad p_0 + 2p - 2q > 0.$$

The choice $p_0 = 0$, which corresponds to Theorem 4, is good only for $b < 5$. For $5 \leq b < b_1$ we need $p_0 \gg 0$ to satisfy the inequality for the origin. Geometrically (see Jansen [16] or the proof of Theorem 19.6.2 in [14]) this means that for $5 \leq b < b_1$, the interior fixed point F_{12}^{12} is contained in the convex hull of the five fixed points $O, F_1^1, F_1^2, F_2^1, F_2^2$ (but the cone D does not yet meet C_γ, the hyperplane through the four points F_i^j, and hence Theorem 7 applies). For $b = b_1$, F_{12}^{12} lies exactly on this hyperplane. For larger values of b, F_{12}^{12} lies in the hull of the F_i^j and F_{12}. So only for $b < 5$, F_{12}^{12} lies outside the convex hull of all boundary fixed points and Theorem 5 yields permanence.

Acknowledgements. I am indebted to J. Reineck for discussions on Conley's theory. This work was partially supported by the Austrian Fonds zur Förderung der wissenschaftlichen Forschung, Project P 5994. Part of it was done during a stay at IIASA, Laxenburg, Austria.

After presenting this lecture again at SISSA in Trieste in February 1988, F. Zanolin showed me a paper by B.M. Garay [10], who had the very same idea of applying the Ura-Kimura theorem and Conley's structure theorem to the question of persistence. I recommend his paper for more details.

References

1 N. P. Bhatia and G. P. Szegö: *Stability Theory of Dynamical Systems.* Grundlehren math. Wissensch. **161**. Berlin - Heidelberg - New York: Springer. 1970.

2 R. Bowen: ω-limit sets for Axiom A diffeomorphisms. J. Diff. Equ. **18**, 333-339 (1975).

3 G. Butler, H. I. Freedman and P. Waltman: Uniformly persistent systems. Proc. Amer. Math. Soc. **96**, 425-430 (1986).

4 G. Butler, P. Waltman: Persistence in dynamical systems. J. Diff. Equ. **63**, 255-263 (1986).

5 C. Conley: *Isolated invariant sets and the Morse index.* CBMS 38. Providence, R.I.: Amer. Math. Soc. 1978.

6 A. Fonda: Uniformly persistent semi-dynamical systems. Proc. Amer. Math. Soc. To appear.

7 H. I. Freedman and J. W.-H. So: Persistence in discrete semi-dynamical systems. Preprint (1987).

8 H. I. Freedman and P. Waltman: Mathematical analysis of some three-species food-chain models. Math. Biosci. **33**, 257-276 (1977).

9 T. C. Gard and T. G. Hallam: Persistence of food webs: I. Lotka-Volterra food chains. Bull. Math. Biol. **41**, 877-891 (1979).

10 B. M. Garay: Uniform persistence and chain recurrence. J. Math. Anal. Appl. To appear.

11 J. Hofbauer: A general cooperation theorem for hypercycles. Monatsh. Math. 91: 233-240 (1981).

12 J. Hofbauer: Heteroclinic cycles on the simplex. Proc. Int. Conf. Nonlinear Oscillations. Budapest 1987.

13 J. Hofbauer and K. Sigmund: Permanence for replicator equations. In: *Dynamical Systems.* Ed. A. B. Kurzhansky and K. Sigmund. Springer Lect. Notes Econ. Math. Systems **287**. 1987.

14 J. Hofbauer and K. Sigmund: *Dynamical Systems and the Theory of Evolution*. Cambridge Univ. Press 1988.

15 V. Hutson: A theorem on average Ljapunov functions. Monatsh. Math. **98**, 267-275 (1984).

16 W. Jansen: A permanence theorem for replicator and Lotka-Volterra systems. J. Math. Biol. **25**, 411-422 (1987).

17 G. Kirlinger: *Permanence of some four-species Lotka-Volterra systems*. Dissertation. Universität Wien. 1987.

18 C. Robinson: Stability theorems and hyperbolicity in dynamical systems. Rocky Mountain J. Math. **7**, 425-434 (1977).

19 P. Schuster, K. Sigmund and R. Wolff: Dynamical systems under constant organization. III. Cooperative and competitive behaviour of hypercycles. J. Diff. Equ. **32**, 357-368 (1979).

20 T. Ura and I. Kimura: Sur le courant exterieur a une region invariante. Theoreme de Bendixson. Comm. Math. Univ. Sanctii Pauli **8**, 23-39 (1960).

Acta Applicandae Mathematicae **14** (1989), 23–35.

Reflections and Calculations on a
Prey-Predator-Patch Problem

O. Diekmann

Centre for Mathematics and Computer Science
Kruislaan 413, 1098 SJ Amsterdam, the Netherlands
&
Institute of Theoretical Biology, University of Leiden
Groenhovenstraat 5, 2311 BT Leiden, the Netherlands

J.A.J. Metz

Institute of Theoretical Biology, University of Leiden
Groenhovenstraat 5, 2311 BT Leiden, the Netherlands

M.W. Sabelis [*]

Department of Population Biology, University of Leiden
P.O. Box 9516, 2300 RA Leiden, the Netherlands

This paper is concerned with models for the interaction of plants, herbivores and their predators. We concentrate on situations in which local colonies of herbivores either over-exploit their host plant or are driven to extinction by predators. Starting from a complicated structured model, in which the local prey and predator density within patches is taken into account, we use time scale arguments to derive a three dimensional system of ordinary differential equations. The simplified system is analysed and the existence of multiple stable steady states is demonstrated.

1980 Mathematics Subject Classification: 92A15
Key Words & Phrases: physiologically structured population models, predator-prey-plant interaction, patch structure, model simplification, time scale arguments, qualitative analysis, multiple stable steady states

1. INTRODUCTION

This paper reports some recent work on a collection of mathematical models for the interaction of phytophages and their natural enemies in an ensemble of local patches of host plants. The key idea is to consider a local colony as an "individual" characterized by the number of prey x, the number of predators y and, possibly, some index for the available food for the prey such as host plant leaf area or biomass. Once the dynamics at the "individual" level are specified one can employ a general methodology (essentially just correct bookkeeping; see Metz & Diekmann, 1986) to derive a "population" model.

The ideal then is to understand the global dynamical behaviour and in particular

---

[*] Present address: Department of Pure and Applied Ecology, University of Amsterdam, Kruislaan 302, 1098 SM Amsterdam, the Netherlands.

how this behaviour is affected by the various ingredients of the (sub)model(s). To attain this ideal for a nonlinear infinite dimensional dynamical system involving many parameters is a next to impossible task. Therefore we have to have recourse to simplifications.

A true understanding of natural phenomena quite often requires a whole spectrum of supplementary models rather than one particular model. We advocate the use of structured models to fill the gap between realistic but complicated simulation models on the one end and qualitative caricatures in terms of ordinary or functional differential equations on the other extreme. The exercise of formulating explicitly a complicated structured model is useful in itself since foggy notions are clarified in the process and questions are identified. As a next step time scale arguments (quasi steady-state assumptions or neglect of delays) or special choices of model ingredients may be employed to derive analytically tractable simplifications. Thus one obtains a coherent *network of models,* and qualitative insights derived from the simplest elements may be used to give direction to numerical experiments on the more intricate elements and to guide the interpretation of the outcomes.

In a recent survey (Diekmann, Metz & Sabelis, 1988) we have illustrated this approach to the modelling of predator-prey interactions in a patchy environment by means of several examples of possible simplifications and the biological conclusions derived from these. In the present more limited paper we concentrate on one rather drastic simplification resulting in a system of three ordinary differential equations which we shall analyse in some detail.

In section 2 we present the structured "master" model while section 3 is devoted to a time scale argument and the resulting simplification. Section 4 deals with the existence, multiplicity and stability of steady states of the three dimensional ode system and, finally, in section 5 the main conclusions are translated into biological terms.

2. MODEL FORMULATION

Consider a herbivorous prey population living scattered over many local patches. New prey colonies are founded by individuals emigrating from existing prey colonies and invading "empty" patches of host plants. Prey colonies come to an end when the host plants are locally over-exploited or when predator invasion has eventually resulted in complete extermination of the prey followed by dispersal of the predators.

Let x denote the number of prey in a given patch. Consider a patch in which only prey are present. We assume that the process of prey colony growth is described by the ordinary differential equation $\frac{dx}{dt} = v(x)$ until the host plant is locally over-exploited or the colony is invaded by a predator. If we assume that all empty patches offer an identical prospect for the prey then the number of prey in a colony which crashes due to host plant over-exploitation is a constant, which we shall call x_{max} (so x_{max} is the exploitable energy of an empty patch expressed in prey equivalents).

Let $Q(t)$ denote the number of potentially invading predators around at time t. Assuming mass action kinetics we let the per colony rate at which prey colonies of size x are invaded be given by $\eta(x)Q(t)$, where the vulnerability η describes how attractive (or, conspicuous) a prey patch of size x is.

Let $n_0(t)$ denote the number of suitable empty patches at time t and let $P(t)$ denote the number of potential prey colonists around at time t, then, again assuming mass action kinetics, the rate at which new prey colonies are founded is given by $\zeta n_0(t)P(t)$, where ζ denotes a reaction constant.

To describe the "population" level we now introduce the density function $n(t,x)$ which is such that the number of patches at time t with prey level between x_1 and x_2 is given by

$$\int_{x_1}^{x_2} n(t,\xi)d\xi.$$

Straightforward bookkeeping arguments (Metz & Diekmann, 1986, p. 15, 92-97, 101) then yield the balance laws

$$\begin{cases} \dfrac{\partial n}{\partial t}(t,x) + \dfrac{\partial}{\partial x}(v(x)n(t,x)) = -\eta(x)Q(t)n(t,x), & 1<x<x_{max}, \\ v(1)n(t,1) = \zeta n_0(t)P(t). \end{cases} \quad (2.1)$$

Any invaded prey patch becomes a (prey-) predator patch. To describe such patches we introduce the number of predators y as another state variable. We assume that the local prey-predator interaction is described by the system of ordinary differential equations

$$\frac{dx}{dt} = g(x,y) \quad , \quad \frac{dy}{dt} = h(x,y).$$

Let the density function $m(t,x,y)$ be such that at time t the number of patches with prey level between x_1 and x_2 and predator level between y_1 and y_2 is given by

$$\int_{x_1}^{x_2} \int_{y_1}^{y_2} m(t,x,y)dydx$$

then our assumptions entail the balance laws

$$\begin{cases} \dfrac{\partial}{\partial t}m(t,x,y) + \dfrac{\partial}{\partial x}(g(x,y)m(t,x,y)) + \dfrac{\partial}{\partial y}(h(x,y)m(t,x,y)) = 0 \\ h(x,1)m(t,x,1) = \eta(x)Q(t)n(t,x) \end{cases} \quad (2.2)$$

A precise description of the domain in the (x,y)-plane in which the differential equation holds requires a submodel for host plant consumption by the prey in order to compute the "resource exhaustion boundary" (see Metz & Diekmann, 1986, p.82 for the simplest possible example). Here we shall neglect this point since the limiting case we are going to consider is chosen such that it becomes irrelevant. The assumption that the predators drive the prey locally to extinction translates into the assumption that the orbits of the prey-predator interaction system connect the invasion boundary $y=1$ with the extermination boundary $x=0$.

Let μ and ν denote the death rates of, respectively, the prey and predator aireal plankton. In accordance with our previous assumptions we describe the dynamics of P and Q by (see Metz & Diekmann, 1986, p. 98-99)

$$\frac{dP}{dt}(t) = x_{max}v(x_{max})n(t,x_{max}) - \mu P(t) \tag{2.3}$$

$$\frac{dQ}{dt}(t) = -\int_1^{y_{max}} yg(o,y)m(t,o,y)dy - \nu Q(t) \tag{2.4}$$

where we have ignored the possible increase of P and Q due to prey and predators dispersing from patches reaching the resource exhaustion boundary in the (x,y)-plane. Concerning the number of empty patches n_0 we shall assume that

$$\frac{dn_0}{dt} = f(n_0) - \zeta n_0 P \tag{2.5}$$

where f is, for example, the familiar logistic function

$$f(n_0) = rn_0(1 - \frac{n_0}{K}).$$

Provided with appropriate initial conditions the equations (2.1) - (2.5) yield a complete dynamical description of the system.

3. INSTANTANEOUS HOST PLANT DESTRUCTION

Suppose the prey exhaust their host plant very quickly compared with the time scale of dispersal, then the founding of a prey colony leads almost instantaneously to the production of new searching prey unless predator invasion precludes over-exploitation in which case the yield consists of predators rather than prey. How do we translate this verbal description of a limiting case into a mathematical simplification of (2.1) - (2.5)?

Solving (2.1) and (2.2) by integration along characteristics (see e.g. Metz & Diekmann, 1986, p. 68-69, 104-105) one can express $n(t,x)$ and $m(t,x,y)$ in past values of n_0, P and Q. Substitution of these expressions into (2.3) - (2.5) then yields a closed system of three delay differential equations. In the limiting case this becomes a system of three ordinary differential equations which describes the system by following the number of empty patches as well as the prey and predator aireal plankton as a function of time, while the rise and annihilation of local colonies are reduced to point events.

To actually calculate the right-hand side of the differential equations we specialize by taking $v(x)=\alpha x, g(x,y)=\alpha x - \beta y$ and $h(x,y)=\gamma y$ (that is, we take exponential prey growth in the absence of predators and a constant functional and numerical response) and let $\alpha,\beta,\gamma\to\infty$. To let predator invasion still be appreciable we have to let $\eta\to\infty$ as well with α and η of the same order. The interpretation suggests to take β and γ of the same order. In order to avoid the complication of patches reaching the resource exhaustion boundary in the (x,y)-plane we let $\frac{\alpha}{\gamma}\to 0$ or, in other words, we assume that the predators reproduce an order of magnitude faster than the prey. In Appendix II of Diekmann, Metz & Sabelis (1988) it is shown that under these assumptions the limiting system of ode's is

$$\frac{dn_0}{dt} = f(n_0) - \zeta n_0 P$$

$$\frac{dP}{dt} = x_{\max} \zeta n_0 P e^{-\omega Q} - \mu P \tag{3.1}$$

$$\frac{dQ}{dt} = \zeta n_0 P h(Q) - \nu Q$$

where by definition

$$\omega = \int_1^{x_{\max}} \frac{\eta(\sigma)}{\alpha \sigma} \, d\sigma$$

$$h(Q) = \frac{\beta}{\gamma} \int_{1+\frac{\gamma}{\beta}}^{1+\frac{\gamma}{\beta} x_{\max}} y d[1 - e^{-Q \int \frac{\eta(\sigma)}{\alpha \sigma} d\sigma}]$$

So the rate of production of prey aireal plankton equals the product of the yield factor $x_{\max}$, the rate of founding of new prey colonies $\zeta n_0 P$ and a reduction factor $\exp(-\omega Q)$ to account for predator invasion. The function h describes how the mean yield of predators per founded prey patch depends on the current predator aireal plankton $Q(t)$. (Note that the yield in predators depends on the size of the prey colony at the moment of invasion while the probability of invasion at some particular size depends on the vulnerability η as well as on Q. The per capita yield $h(Q)/Q$ is monotone decreasing.)

Specializing still further we take $\eta(x) = \alpha \delta x$ which means that we assume that the probability of predator invasion is proportional to the prey colony size. Then

$$\omega = \delta(x_{\max} - 1) \tag{3.2}$$

$$h(Q) = (\frac{\beta}{\gamma} + x_{\max})(1 - e^{-\omega Q} + \rho(\frac{1 - e^{-\omega Q}}{\omega Q} - 1)), \quad Q > 0 \tag{3.3}$$

and $h(0) = \lim_{Q \downarrow 0} h(Q) = 0$. Here

$$\rho = \frac{x_{\max} - 1}{x_{\max} + \beta/\gamma} \tag{3.4}$$

The graph of h is sketched in Figure 1. Note that h decreases for large values of Q! One can prove analytically that h has exactly one maximum for positive Q.

In the next section we shall analyse the system (3.1) with the empty patch production function f given by

$$f(n_0) = r n_0 (1 - \frac{n_0}{K}) \tag{3.5}$$

The relation between h and Q for $\omega = 1$, $\frac{\beta}{\gamma} + x_{\max} = 1$ and various values of ρ

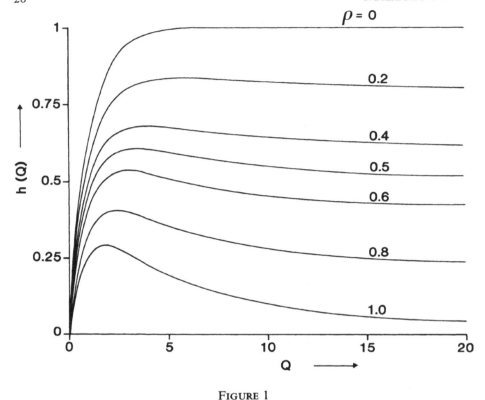

4. STABILITY AND BIFURCATION

We begin our analysis of (3.1) (with ω, h, ρ and f given by (3.2) - (3.5)) by performing a scaling. Define

$$\begin{cases} u(t) = \dfrac{x_{max}\zeta}{\mu} \, n_0(\dfrac{t}{\mu}) \\[2mm] v(t) = \dfrac{\zeta}{\mu} \, P(\dfrac{t}{\mu}) \\[2mm] w(t) = \omega Q(\dfrac{t}{\mu}) \end{cases} \tag{4.1}$$

$$\begin{cases} a = \dfrac{\mu\omega}{x_{max}\zeta}(\dfrac{\beta}{\gamma} + x_{max}) & b = \dfrac{r}{\mu} \\[2mm] c = \dfrac{x_{max}\zeta K}{\mu} & d = \dfrac{\nu}{\mu} \end{cases} \tag{4.2}$$

then the system (3.1) can be rewritten as

$$
\begin{cases}
\dfrac{du}{dt} = bu(1-\dfrac{u}{c})-uv \\[2mm]
\dfrac{dv}{dt} = uve^{-w}-v \\[2mm]
\dfrac{dw}{dt} = auv(1-e^{-w}+\rho(\dfrac{1-e^{-w}}{w}-1)) - dw
\end{cases}
\tag{4.3}
$$

The equilibria of this scaled system are:

i) $u=v=w=0$ (no empty patches; no herbivores; no predators)

ii) $u=c,v=w=0$ (empty patches at carrying capacity; no herbivores; no pre-dators)

iii) $u=1, v=b(1-c^{-1}), w=0$ (requires $c>1$; the density of empty patches is completely set by the "predation" pressure of the herbivores; no predators)

iv) $v=b(1-\dfrac{u}{c}), w=lnu,(1-\dfrac{u}{c})(u-1+\rho\ (\dfrac{u-1}{lnu}-1))=\dfrac{d}{ab}lnu$ (steady state with three trophic levels present; no explicit expression for u)

The steady state (i) is unstable for all $b>0$, whereas (ii) is stable for $0<c<1$ and unstable for $c>1$. Linearization about the steady state (iii) yields the matrix

$$
\begin{bmatrix}
-\dfrac{b}{c} & -1 & 0 \\[3mm]
b(1-\dfrac{1}{c}) & 0 & -b(1-\dfrac{1}{c}) \\[3mm]
0 & 0 & ab(1-\dfrac{1}{c})(1-\dfrac{1}{2}\rho)-d
\end{bmatrix}
$$

which has eigenvalues

$$
\lambda_1 = ab(1-\dfrac{1}{c})(1-\dfrac{1}{2}\rho)-d
$$

$$
\lambda_{2,3} = -\dfrac{b}{2c}\pm\sqrt{\dfrac{b^2}{4c^2}-b(1-\dfrac{1}{c})}.
$$

Clearly $Re\lambda_{2,3}<0$ for $c>1$, while $\lambda_1<0$ if and only if

$$
c(1-\dfrac{1}{2}\rho-\theta)<1-\dfrac{1}{2}\rho
$$

where by definition

$$
\theta = \dfrac{d}{ab}.
\tag{4.4}
$$

The definition of ρ (see 3.4) implies that $0 \leqslant \rho < 1$ (we will make this hypothesis throughout the rest of the paper), so $1 - \frac{1}{2}\rho > 0$. Hence the steady state (iii) is stable for all values of $c > 1$ if $1 - \frac{1}{2}\rho - \theta < 0$, whereas it is stable for

$$1 < c < \frac{1 - \frac{1}{2}\rho}{1 - \frac{1}{2}\rho - \theta}$$ and unstable for larger values of c if $1 - \frac{1}{2}\rho - \theta > 0$.

We now turn our attention to the steady state (iv) which is only implicitly defined. The easiest way to proceed seems to change our point of view and consider c as a function of u:

$$c(u) = \frac{u(u - 1 + \rho \dfrac{u - 1 - u\ln u}{\ln u})}{u - 1 + \rho \dfrac{u - 1 - u\ln u}{\ln u} - \theta \ln u} \tag{4.5}$$

We first investigate where and how this curve in the (c, u)-plane intersects the line $u = 1$ corresponding to the steady state (iii). If we put $u = 1 + \epsilon$ and make a Taylor expansion with respect to ϵ we find

$$c(1 + \epsilon) = \frac{1 - \frac{1}{2}\rho}{1 - \frac{1}{2}\rho - \theta} + \epsilon \frac{(1 - \frac{7}{12}\rho)(1 - \frac{1}{2}\rho - \theta) - (1 - \frac{1}{2}\rho)(\frac{1}{2}\theta - \frac{1}{12}\rho)}{(1 - \frac{1}{2}\rho - \theta)^2} + h.o.t. \tag{4.6}$$

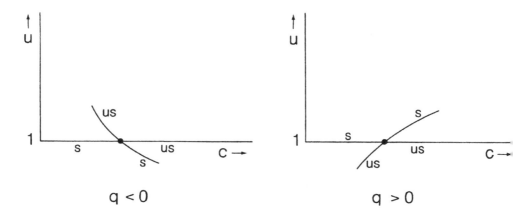

FIGURE 2

Local bifurcation diagram. q is the coefficient of ϵ in (4.6).
s means stable and us unstable.

Thus we find, as to be expected, that for $1 - \frac{1}{2}\rho - \theta < 0$ no intersection occurs in the positive quadrant while for $1 - \frac{1}{2}\rho - \theta > 0$ intersection occurs exactly at the point where steady state (iii) loses its stability. Define $q = q(\rho, \theta)$ to be the coefficient of ϵ

in the expansion then the local configuration is as depicted in Figure 2, where the stability assertions about steady state (iv) are based on the general principle of the exchange of stability in a bifurcation point (see Metz & Diekmann (1986) VI. 1.2 and the references given there). Note that the branch with $u<1$ is biologically meaningless since $w<0$. The direction of bifurcation changes for $q=0$ which corresponds to $\theta=3(2-\rho)^2/(18-10\rho)$ (see Figure 3).

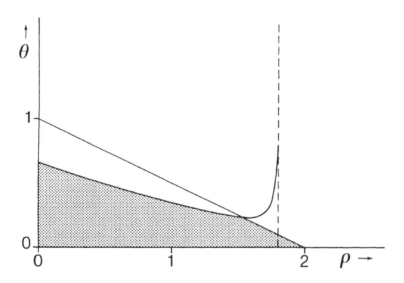

FIGURE 3
The set in the (ρ,θ)-plane such that bifurcation occurs is
the triangle $\rho,\theta\geqslant0$, $\theta<1-\frac{1}{2}\rho$.
Within the hatched area bifurcation is supercritical and
steady state (iv) is stable near the bifurcation point, whereas in the
unhatched area of the triangle we have subcritical bifurcation and
steady state (iv) is unstable near the bifurcation point.

Let us now try to obtain information about the *global* aspects of the bifurcation diagram. We have to address two partly related problems: what is the shape of $c(u)$ and what is the stability character. We begin by investigating the shape.

Since the numerator of $c(u)$ is always positive for $u>1$ and $0\leqslant\rho<1$ the following lemma shows that $c(u)$ has precisely one positive branch for $u>1$.

LEMMA 4.1. *The denominator of $c(u)$ as defined in (4.5) has no zero's for $u>1$ when $1-\frac{1}{2}\rho-\theta>0$ and precisely one zero when $1-\frac{1}{2}\rho-\theta<0$.*

PROOF. We define

$$\psi(z) = (1-\rho)ze^z + \rho e^z - \rho - z - \theta z^2$$

and note that the zero's of ψ with $z>0$ are in one to one correspondence with the zero's of the denominator via $u=e^z$. Then

$$\psi'(z) = (1-\rho)ze^z + e^z - 1 - 2\theta z$$

$$\psi''(z) = (1-\rho)ze^z + (2-\rho)e^z - 2\theta$$

$$\psi'''(z) = (1-\rho)ze^z + (3-2\rho)e^z$$

and consequently $\psi'''(z)>0$ for $z>0$. For $z\sim 0$ we have

$$\psi(z) \sim (1-\tfrac{1}{2}\rho-\theta)z^2$$

and for $z\to\infty$ we have $\psi(z)\to\infty$. Suppose $1-\tfrac{1}{2}\rho-\theta>0$. If ψ has one positive zero it has to have at least two positive zero's. Since ψ is increasing for small positive z the function ψ' has to have at least two positive zero's as well. Consequently ψ'' has at least one positive zero. But $\psi''(0)=1-\tfrac{1}{2}\rho-\theta>0$ and $\psi'''(z)>0$ for $z>0$, so ψ'' cannot have a positive zero.

Next consider the case that $1-\tfrac{1}{2}\rho-\theta<0$. Then ψ has an odd number of positive zero's. Assume this number is three or more. Since ψ is decreasing for small positive z the function ψ' has to have at least three positive zero's. Applying the same argument to ψ' we deduce that ψ'' has at least three positive zero's. However, since $\psi'''>0$ we know that ψ'' has only one positive zero. We conclude that ψ cannot have more than one positive zero. $\square$

In principle the branch could have several wiggles and therefore we could have, for specific values of c, even more than two steady states with three trophic levels occupied. In the special case $\rho=0$ we can exclude the possibility of wiggles.

LEMMA 4.2. *Let* $c(u)$ *be defined by* (4.5). *For* $\rho=0$ *and any* $c>0$ *the set* $\{u>1\,|\,c(u)=c\}$ *contains at most two elements.*

PROOF. Define for fixed c the function F by

$$F(u) = u(u-1)-c(u-1-\theta\ln u).$$

Clearly there is a one to one correspondence between the zero's of F with $u>1$ and the set $\{u>1\,|\,c(u)=c\}$. Note that $F'''(u)=\dfrac{2c\theta}{u^3}>0$ for $u>0$ and that $F(1)=0, F(\infty)=\infty$ and $F'(1)=1-c(1-\theta)$. Employing the same arguments as in the proof of Lemma 4.1 it then follows that F has precisely one zero for $u>1$ if $F'(1)<0$ whereas F has either no or two zero's for $u>1$ when $F'(1)>0$. Note that $F'(1)=0$ exactly at the bifurcation point $c=(1-\theta)^{-1}$. $\square$

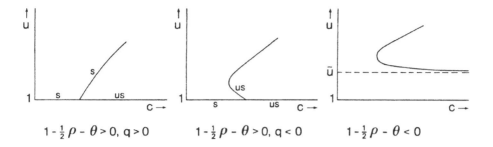

$$1 - \tfrac{1}{2}\rho - \theta > 0, \; q > 0 \qquad 1 - \tfrac{1}{2}\rho - \theta > 0, \; q < 0 \qquad 1 - \tfrac{1}{2}\rho - \theta < 0$$

FIGURE 4
Global bifurcation diagram (rigorously verified only for $\rho=0$).
s means stable and us unstable.

We have now verified the pictures of Figure 4 for the special case $\rho=0$. We conjec-ture that they are correct for $0<\rho<1$ as well, but for this we only have some numerical evidence obtained by solving $\dfrac{d}{du}c(u)=0$ for θ and plotting θ so defined as a function of u, for many values of $\rho\in(0,1)$. The results suggest that for fixed ρ and θ there is at most one turning point (i.e. a point where $\dfrac{d}{du}c(u)=0$). We tried to exclude the "birth" of a turning point analytically by looking at second deriva-tives but even though one can reduce the problem to a quadratic equation in ρ (with coefficients depending in a complicated way on u) we did not manage to find a proof.

Next we turn our attention to the stability problem. The Jacobi matrix at a steady state $c(u)=c$ is given by

$$
\begin{pmatrix}
-\dfrac{b}{c}u & -u & 0 \\[2mm]
\dfrac{b}{u}(1-\dfrac{u}{c}) & 0 & -b(1-\dfrac{u}{c}) \\[2mm]
\dfrac{d}{u}lnu & \dfrac{dlnu}{b(1-\dfrac{u}{c})} & \dfrac{dlnu\,g'(lnu)}{g(lnu)}
\end{pmatrix}
$$

where by definition

$$g(w) = \frac{1-e^{-w}}{w} + \rho\frac{1-e^{-w}-w}{w^2}.$$

The characteristic equation reads

$$\lambda^3 + a_1\lambda^2 + a_2\lambda + a_3 = 0$$

where

$$a_1 = \frac{b}{c}u - dlnu\frac{g'(lnu)}{g(lnu)}$$

$$a_2 = -\frac{bd}{c}ulnu\frac{g'(lnu)}{g(lnu)} + dlnu + b(1-\frac{u}{c})$$

$$a_3 = bd(1-\frac{u}{c})lnu\left[\frac{u}{c-u}-1-\frac{g'(lnu)}{g(lnu)}\right].$$

Note that $\lambda=0$ is a root iff $a_3=0$ and that $a_3=0$ iff either $u=1$ or $\frac{g'(lnu)}{g(lnu)}=1-\frac{u}{c-u}$. The first possibility corresponds to the bifurcation from the $(u=1,v=b(1-\frac{1}{c}),w=0)$ branch and the second, as we are going to show, to the turning point(s) of the $c(u)$ branch. Indeed

$$c(u) = \frac{u^2g(lnu)}{ug(lnu)-\theta}$$

and therefore

$$(ug(lnu)-\theta)^2c'(u) = \frac{u^2g^2(lnu)}{c}\{(u-c)\frac{g'(lnu)}{g(lnu)} + 2u-c\}.$$

So $c'(u)>0$ iff

$$\frac{g'(lnu)}{g(lnu)} < \frac{c-2u}{u-c} = -1 + \frac{u}{c-u}.$$

It follows that for $u>1$ $a_3>0$ iff $c'(u)>0$. In other words, at turning points of the $c(u)$ branch a real root changes from the left half plane (when $c'(u)>0$; this follows from the Routh-Hurwitz criteria, see below) to the right half plane (when $c'(u)<0$).

The stability of the steady state may also change by a pair of complex conjugated roots crossing the imaginary axis. Note that the characteristic equation has roots exactly on the imaginary axis iff $a_2>0$ and $a_1a_2=a_3$. Finally, recall that the Routh-Hurwitz criteria for stability are

$$a_1>0, \quad a_3>0 \quad \text{and} \quad a_1a_2>a_3$$

(and that this necessitates a_2 to be positive as well).

Armed with the above observations we will now show that the stability problem does not admit a simple solution. Clearly $c'(u)<0$ and $u>1$ imply instability but the tempting conjecture that $c'(u)>0$ and $u>1$ imply stability is false. A key point is that the $c(u)$ branch depends only on the compound parameter $\theta=\frac{d}{ab}$ whereas the coefficients a_1,a_2 and a_3 of the characteristic polynomial depend on b and d individually. If we let a and d tend to zero while keeping b,θ,ρ,u and c constant the expressions for a_1,a_2 and a_3 show that both a_1 and a_2 are positive and bounded away from zero while a_3 tends to zero from above when $c'(u)>0$. It follows that the Routh-Hurwitz stability criteria are satisfied for small a and d.

If, on the other hand, we let a and d tend to infinity while keeping all the other quantities constant then (note that $g'<0$) a_1 becomes negative and the steady state is unstable. In fact a pair of eigenvalues must cross the imaginary axis so that Hopf bifurcation theory implies the existence of a branch of periodic solutions of the system of ordinary differential equations.

We conclude that any given steady state on a part of the $c(u)$ branch with $c'(u)>0$ may either be stable or unstable, depending on the precise values of the parameters a,b, and d. For small a and d and $1-\frac{1}{2}\rho-\theta<0$ or $1-\frac{1}{2}\rho-\theta>0, q<0$ we have at least two *stable* steady states. It seems very likely that for large a and d and $1-\frac{1}{2}\rho-\theta<0$ or $1-\frac{1}{2}\rho-\theta>0, q<0$ a stable steady state and a stable limit cycle coexist.

5. BIOLOGICAL CONCLUSIONS

The limiting ode system admits two stable steady states (as well as one unstable steady coexistence state) in a large domain of parameter space. In one of the stable steady states the predators are absent and the herbivores keep the number of "empty" host plant patches n_0 far below the carrying capacity K. If one tries to apply biological control by introducing a small number of predators the stability of this steady state prevents success. However, the introduction of a large number of predators may bring the system into the other stable steady-state in which the plants are almost at the carrying capacity (note that $c(u)\sim c$ for $u\to\infty$) while the herbivores are kept at a low level by the predators. There also exist regions in parameter space in which the latter steady state is unstable and, presumably, stable oscillations around this steady state exist.

In terms of the original parameters we have

$$\theta = \frac{\nu x_{max}\zeta}{\mu r\delta(x_{max}-1)(\frac{\beta}{\gamma}+x_{max})}, \qquad \rho = \frac{x_{max}-1}{x_{max}+\beta/\gamma}.$$

Hence $\theta>1-\frac{1}{2}\rho$ if and only if

$$x_{max} < \frac{\nu\zeta}{\mu r\delta} - \frac{\beta}{\gamma} + \sqrt{(\frac{\nu\zeta}{\mu r\delta} - \frac{\beta}{\gamma})^2 + 1 + \frac{2\beta}{\gamma}}.$$

So if we think of situations with equal total exploitable host plant biomass $x_{max}K$ then one can expect multiple stable steady states when there are many small patches and a single stable steady state when the patches are large but few.

6. REFERENCES

O. DIEKMANN, J.A.J. METZ & M.W. SABELIS (1988). *Mathematical models of predator-prey-plant interactions in a patchy environment,* to appear in a special issue of Experimental and Applied Acarology edited by M.W. Sabelis, A.R.M. Janssen and W. Helle.

J.A.J. METZ & O. DIEKMANN (eds.) (1986). *The Dynamics of Physiologically Structured Populations.* Springer Lecture Notes in Biomathematics **68.**

Acta Applicandae Mathematicae 14 (1989), 37–47.
© 1989 *by IIASA*.

A Homotopy Technique for a Linear Generalization

of Volterra Models

Edoardo Beretta

Istituto Matematico "Ulisse Dini"
Università di Firenze
Viale Morgagni 67/A
I-50134 Firenze, Italy

AMS Subject Classification (1980): 92A17
Key words: Volterra equations, Ljapunov functions, homotopy path, diffusion

1. Introduction

The classical Lotka-Volterra models from population dynamics have the structure of
the system of O.D.E.

$$\frac{dx_i}{dt} = x_i \left(e_i + \sum_{j \in N} a_{ij} x_j \right) \quad , \quad i \in N \quad , \tag{1.1}$$

where $N = \{1,2,\ldots,n\}$ is the set of all the indices of the variables , e_i , a_{ij} ,
$i,j \in N$ are suitable real parameters and $x_i = x_i(t)$ represents the density or the
biomass of i-th species at time t.

If we denote by:

$$e = col(e_1,\ldots,e_n) \in R^n,$$
$$x = col(x_1,\ldots,x_n) \in R^n_{+o} \quad , \quad R^n_{+o} := \{ x \in R^n : x_i \geq 0 \ , \ i \in N \} \tag{1.2}$$
$$A = (a_{ij})_{i,j \in N}$$

system (1.1) can be set in the matrix form:

$$\frac{dx_i}{dt} = x_i(e + Ax)_i \quad , \quad i \in N \quad , \tag{1.3}$$

where any solution $x = x(t;t_o,x_o)$ of (1.3) will remain in R^n_{+o} for all times
$t \in [t_o,+\infty)$ provided that at $t=t_o$ $x_o \in R^n_{+o}$.
By the linear Complementary Problem (see: Berman and Plemmons, 1979), Takeuchi and Adachi (1980) proved that $-A \in S_w$ is a sufficient condition for assuring the existence,
for each given vector $e \in R^n$, of a nonnegative equilibrium $x^{::}$ of (1.3) which is
globally asymptotically stable in R^n_I ,

$$R^n_I : = \{x \in R^n : x_i > 0 \ , \ i \in N-I\} \subseteq R^n_{+o} \tag{1.4}$$

where I is the subset of N of all the indices such that $x^{::}_i = 0$.
In a temptative of including most part of the known deterministic epidemic models in
a general class of O.D.E. Beretta and Capasso (1986) proposed the following " linear

generalization of Lotka-Volterra models ":

$$\frac{dx_i}{dt} = x_i e_i + x_i \sum_{j \in N} a_{ij} x_j + c_i + \sum_{j \in N}' b_{ij} x_j \quad , \quad i \in N \tag{1.5}$$

where " $x_i e_i$ " includes the term " $b_{ii} x_i$ " .
The positive invariance of R_{+o}^n requires that in (1.5) the following holds:

$$c_i \geq 0 \quad , \quad i \in N \quad ; \quad b_{ij} \geq 0 \ (b_{ii} = 0) \quad , \quad i,j \in N \tag{1.6}$$

By introducing the further nomenclature:

$$c = \text{col} \ (c_1, \ldots, c_n) \in R_{+o}^n \quad ,$$
$$B = (b_{ij})_{i,j \in N} \in R_{+o}^{n \times n} \tag{1.7}$$

we define the nonnegative linear vector on R_{+o}^n :

$$b(x) = c + Bx \quad , \quad b(x) \geq 0 \quad \text{for all} \quad x \in R_{+o}^n \tag{1.8}$$

and we can set (1.5) in the matrix form:

$$\frac{dx_i}{dt} = x_i (e + Ax)_i + b_i (x) \quad , \quad i \in N \quad . \tag{1.9}$$

On the right side of (1.9) the first term is the usual Lotka-Volterra Part.
Let be $x^{\ast} \geq 0$, $x^{\ast} \in R_{+o}$ a nonnegative equilibrium of (1.9) and $I \subseteq N$ the subset
of all the indices at which $x^{\ast}$ has vanishing components.
Then, the scalar Liapunov function $V: R_I^n \to R_{+o}$ defined as follows:

$$V(x) = \sum_{i \in N-I} w_i \left(x_i - x_i^{\ast} - x_i^{\ast} \ln \frac{x_i}{x_i^{\ast}} \right) + \sum_{i \in I} w_i x_i \quad , \quad w_i \in R_+ \quad , \quad i \in N \tag{1.10}$$

has time derivative that along the trajectories of (1.9) in R_I^n results:

$$\dot{V}(x) \Big|_{R_I^n} = (x - x^{\ast})^T \tilde{W A}(x - x^{\ast}) - \sum_{i \in N-I} \frac{w_i b_i(x)(x_i - x_i^{\ast})^2}{x_i x_i^{\ast}} + \sum_{i \in I} w_i x_i (e + Ax^{\ast})_i , \tag{1.11}$$

where $W = \text{diag}(w_1, \ldots, w_n)$, $w_i \in R_+$, and $\tilde{A}$ is the real constant matrix obtained from
the matrix A in (1.9) according to

$$\tilde{A} = (\tilde{a}_{ij})_{i,j \in N} \quad : \quad \tilde{a}_{ij} = \begin{cases} a_{ij} + \dfrac{b_{ij}}{x_i^{\ast}} \ , \ i \in N-I \ , \ j \in N \\ a_{ij} \quad \text{otherwise} \ . \end{cases} \tag{1.12}$$

We may ask how to derive a sufficient condition for the existence of a nonnegative
equilibrium $x^{\ast}$ of (1.9) globally asymptotically stable in the suitable subspace
R_I^n , similar to that obtained by Takeuchi and Adachi (1980) for Lotka-Volterra models,

when the Linear Complementary   Problem cannot be applied to (1.9) because of the pre
sence of the non vanishing and nonnegative linear vector b(x).

A possible approach is that by a suitable homotopy function which can be defined by
consideration of the structure of the time derivative (1.11) of the Liapunov function
and of the structure of the vector field in (1.9).

2. The simplest case

The simplest case is obtained from (1.9) when vector b(x) in (1.8) has only some
constant components and the others are identically vanishing, i.e.:

$$\frac{dx_i}{dt} = x_i(e + Ax)_i + c_i \quad , \quad i \in N \tag{2.1}$$

where $c_i > 0$ for all $i \in J \subseteq N$, $J \neq \emptyset$, $c_i = 0$ for all $i \in N-J$. The constant terms
c_i in (2.1) may have the biological meaning of " constant currents of immigration "
or that of " prey shelters from predation ".

Let $x^{::}$ be a nonnegative equilibrium of (2.1) and $I \subseteq N-J$ be the subset of all the
indices at which $x_i^{::} = 0$. Let us observe that if $c_i > 0$ for all $i \in N$ then $I = \emptyset$
and $x^{::}$ is a positive equilibrium of (2.1).

Furthermore, since $b_{ij} = 0$ for all $i,j \in N$, from (1.12) it follows that (1.11) can
be rewritten as:

$$\dot{V}(x)\Big|_{R_I^n} = (x - x^{::})^T WA(x - x^{::}) - \sum_{i \in N-I} \frac{w_i c_i (x_i - x_i^{::})^2}{x_i x_i^{::}} + \sum_{i \in I} w_i x_i (e + Ax^{::})_i. \tag{2.2}$$

Hence (2.1) and (2.2) suggest to consider the homotopy function $H: R_{+o}^n \times T \to R^n$,
$T = [0,1]$, with components given by:

$$\begin{aligned}
H_i(x,\xi) &= x_i(e + Ax)_i + \xi c_i \quad , \quad i \in J \\
H_i(x,\xi) &= x_i(e + Ax)_i + \xi(1 - \xi)d_i \quad , \quad d_i \in R_+ \quad , \quad i \in N-J \quad ,
\end{aligned} \tag{2.3}$$

$\xi \in T$. Let us observe that for $\xi = 1$ we get the vector field of (2.1) and when $\xi = 0$
we obtain the vector field of the standard Lotka-Volterra models. We can prove the
following:

THEOREM 2.1. If in (2.1) $-A \in S_w$ then a nonnegative equilibrium $x^{::}$ of (2.1) exists
at which $(e + Ax^{::})_i \leq 0$ for all $i \in N$.

Proof. Let $H = H(x,\xi)$ be the homotopy function (2.3) and let us consider the asso
ciated system of O.D.E.:

$$\dot{x} = H(x,\xi) \quad , \quad x \in R^n \quad , \quad \xi \in T \quad . \tag{2.4}$$

We observe that R_{+o}^n is positively invariant for the solutions of (2.4). As far as the boundedness of the solutions of (2.4) is concerned we introduce the scalar function $S:R_{+o}^n \to R_{+o}$:

$$S := \sum_{i \in N} w_i x_i \quad , \quad w_i \in R_+ \quad , \quad i \in N \quad , \tag{2.5}$$

and we consider $\overset{\cdot}{S} + \varepsilon S$ for some $\varepsilon \in R_+$ along the trajectories of (2.4):

$$\overset{\cdot}{S} + \varepsilon S = x^T WAx + \sum_{i \in N} [x_i (e_i + \varepsilon) + \sigma_i (\xi)] w_i \tag{2.6}$$

where by $\sigma_i(\xi)$ we define the nonnegative constants:

$$\sigma_i(\xi) := \xi c_i \quad , \quad i \in J \quad ; \quad \sigma_i(\xi) := \xi(1 - \xi)d_i \quad , \quad d_i \in R_+ \quad , \quad i \in N-J \quad ; \quad \xi \in T \quad . \tag{2.7}$$

By the hypothesis $-A \in S_w$, i.e. the real symmetric matrix $\frac{1}{2} (WA + A^T W)$ is negative definite. Let be λ_i , $i \in N$ its real negative eigenvalues and let us denote by

$$K = \min_{i \in N} \{|\lambda_i|\} \quad . \tag{2.7}$$

By consideration of (2.7) in (2.6) we obtain that

$$\overset{\cdot}{S} + \varepsilon S \leq \sum_{i \in N} \{-Kx_i^2 + w_i (e_i + \varepsilon)x_i + w_i \sigma_i(\xi)\} \quad , \tag{2.8}$$

i.e., for each $\xi \in T$ and $\varepsilon \in R_+$ there exist a constant $C_\xi > 0 : \overset{\cdot}{S} + \varepsilon S < C_\xi$. Therefore, any solution of (2.4) with initial conditions $x_o \in R_{+o}^n$ such that

$$S_o = \sum_{i \in N} w_i x_{io} < \frac{C_\xi}{\varepsilon} + R \quad , \quad R \geq 0$$

will satisfy the bounds

$$0 \leq S(t) < \frac{C_\xi}{\varepsilon} + \frac{R}{\exp(\varepsilon t)} \quad , \quad t \in [0, +\infty) \quad . \tag{2.9}$$

Let us introduce the positive constant

$$L_\xi := \frac{C_\xi}{\varepsilon} + R \quad . \tag{2.10}$$

The compact subset of R_{+o}^n

$$\Omega_{L_\xi} := \left\{ x \in R_{+o}^n : S \leq L_\xi \right\} \tag{2.11}$$

is positively invariant. Hence (2.4) has a fixed point $x = x(\xi)$ which (as $t \nearrow +\infty$) belongs to:

$$\overline{\Omega}_\xi := \left\{ x \in R_{+o}^n : S \leq \frac{C_\xi}{\varepsilon} \right\} \quad . \tag{2.12}$$

Since for each $\xi \in \overset{o}{T} = (0,1)$ we have $\sigma_i(\xi) > 0$ for all $i \in N$, and the homotopy function is

$$H_i(x,\xi) = x_i(e + Ax)_i + \sigma_i(\xi) \quad , \quad i \in N \tag{2.13}$$

we conclude that:

" if $-A \in S_w$, for each $\xi \in \overset{o}{T}$ there exist a positive solution $x = x(\xi)$ of $H(x,\xi) = 0$

at which

$$(e + Ax(\xi))_i < 0 \quad \text{for all} \quad i \in N .$$

Furthermore, $x = x(\xi) \in \overset{o}{\overline{\Omega}_\xi}$ (i.e. the interior of $\overline{\Omega}_\xi$) ."

Let be

$$D := \left\{ x \in R^n_{+o} : S \leq \max_{\xi \in T} \frac{C_\xi}{\epsilon} \right\} . \tag{2.14}$$

Then $\overline{\Omega}_\xi \subseteq D$ for all $\xi \in T$. Accordingly, we may restrict the homotopy on $D \times T$, i.e. $H : D \times T \to R^n$ and we can state that:

(H-1): if $-A \in S_w$, there exist a compact subset $D \subseteq R^n_{+o} : H(x,\xi) = 0$ has a solution

$x = x(\xi) \in D$ for each $\xi \in T$. If $\xi \in \overset{o}{T} = (0,1)$, $x = x(\xi) \in \overset{o}{D}$ and

$$(e + Ax(\xi))_i < 0 \quad , \quad i \in N . \tag{2.15}$$

Let us introduce

$$\ker H = \{(x,\xi) \in D \times T : H(x,\xi) = 0\} \tag{2.16}$$

$$\overset{o}{\widehat{\ker H}} = \{(x,\xi) \in D \times \overset{o}{T} : H(x,\xi) = 0\} \tag{2.17}$$

Owing to (H-1) $\overset{o}{\widehat{\ker H}} \neq \emptyset$. The homotopy's Jacobian matrix of x-variables is:

$$H'_x(x,\xi) = \text{diag} ((e + Ax)_1,\ldots,(e + Ax)_n) + \text{diag} (x_1,\ldots,x_n)A . \tag{2.18}$$

Because of (H-1), in $\overset{o}{\widehat{\ker H}}$ the homotopy's solution $x = x(\xi)$ is positive and such

that inequalities (2.15) hold true. Therefore, in $\overset{o}{\widehat{\ker H}}$ we can define the diagonal

positive matrix

$$\tilde{W} = \text{diag} \left(\frac{w_1}{x_1(\xi)} ,\ldots, \frac{w_n}{x_n(\xi)} \right) \quad , \quad w_i \in R_+ \tag{2.19}$$

by which we obtain:

$$-\left[\tilde{W} H'_x(x,\xi) + H'_x(x,\xi)^T \tilde{W} \right] = \tilde{W} \text{diag} \left(-(e + Ax(\xi))_1,\ldots,-(e + Ax(\xi))_n \right) +$$

$$- \left[WA + A^T W \right] , \quad W = \text{diag} (w_1,\ldots,w_n) . \tag{2.20}$$

Since $-A \in S_w$, (2.20) implies that in $\overset{o}{\widehat{\ker H}}$ is $-H'_x \in S_w$, i.e. $-H'_x \in P$. By the definition of a P-matrix (Berman and Plemmons, 1979) we conclude that

$$\det \left[-H'_x(x,\xi) \right] > 0 \quad \text{in } \overset{o}{\widehat{\text{ker } H}} \quad . \tag{2.21}$$

The following holds:

(H-2): the homotopy is of full rank in $\overset{o}{\widehat{\text{ker } H}}$, and $\overset{o}{\widehat{\text{ker } H}}$ consists only of continu̲
ously differentiable homotopy paths traversed with

$$\frac{d\xi}{d\rho} = (-1)^{n+1} q(\rho) \det H'_x(x,\xi) \quad . \tag{2.22}$$

In (2.22) $q:R^1 \to R^1$ is an arbitrary continuous function of " ρ " such that $q(\rho) \neq 0$. " ρ " is the distance moved along the path from any initial given solution $(\overline{x} = x(\overline{\xi}), \overline{\xi}) \in \overset{o}{\widehat{\text{ker } H}}$. (Garcia and Zangwill, 1981).
We choose $q(\rho) = (-1)^n$ in order that

$$\frac{d\xi}{d\rho} = \det \left[-H'_x(x,\xi) \right] > 0 \quad \text{in } \overset{o}{\widehat{\text{ker } H}} \quad . \tag{2.23}$$

Thanks to (2.23), any homotopy path is traversed with $\xi = \xi(\rho)$ which is a monotone increasing function of ρ , and thanks to (H-1) the homotopy path cannot reach the boundaries of D as ξ varies in $\overset{o}{\uparrow}$. Therefore, it must reach or tend (as $\rho \nearrow +\infty$) to the homotopy's solution $(x^*, \xi = 1)$ which is a nonnegative equilibrium of (2.1). Since along any homotopy path the inequalities (2.15) hold true, by continuity at $(x^*, \xi = 1)$ we obtain that:

$$(e + Ax^*)_i \leq 0 \quad , \quad i \in N \quad . \tag{2.24}$$

The proof is complete.

Let us observe that, if in (2.22) we choose $q(\rho) = (-1)^{n+1}$, we obtain

$$\frac{d\xi}{d\rho} = \det \left[H'_x(x,\xi) \right] < 0 \quad \text{in } \overset{o}{\widehat{\text{ker } H}} \quad . \tag{2.25}$$

Therefore, any homotopy path is traversed with $\xi = \xi(\rho)$ which is a monotone decre̲
asing function of " ρ " and it must reach or tend (as $\rho \nearrow +\infty$) to the homotopy's solu̲
tion $(x^*, \xi = 0)$ of $H(x,\xi = 0) = 0$, at which $(e + Ax^*)_i \leq 0$, $i \in N$.
Because of the structure (2.3) of the homotopy function, $H = H(x,\xi = 0)$ is the vector
field for the usual Lotka-Volterra models and $(x^*,\xi = 0)$ is a nonnegative equilibrium
at which the properties of the Linear Complementary Problem hold:

$$x^*_i(e + Ax^*)_i = 0 \quad , \quad (e + Ax^*)_i \leq 0 \quad , \quad i \in N \quad . \tag{2.26}$$

Therefore, by the homotopy technique we regain the Linear Complementary Problem pro
perties when $\xi = 0$.

The following is a consequence of Theorem 2.1:

Theorem 2.2. If $-A \in S_w$, and vector e in (2.1) is such that $e_i > 0$ for at least
one $i \in N$, the system (2.1) has a nonnegative (and non-trivial) equilibrium point
$x^{::}$ which is globally asymptotically stable in R_I^n , where I is the subset of all
indices at which $x_i^{::} = 0$.

Proof.

If $-A \in S_w$, Theorem 2.1 implies the existence of a nonnegative equilibrium $x^{::}$ of
(2.1) at which $(e + Ax^{::})_i \leq 0$ $i \in N$. If $e_i > 0$ for some index $i \in N$, $x^{::}$ cannot
be the trivial equilibrium $x^{::} = 0$ otherwise we get a contradiction:

$$\text{for some } i \in N : \quad (e + Ax^{::})_i = e_i > 0 \quad . \tag{2.27}$$

Since in R_I^n the time derivative of the Liapunov function (1.10) is given by (2.2),
the assumption $-A \in S_w$ together with the property $(e + Ax^{::})_i \leq 0$, $i \in N$ is suffi
cient to ensure the global asymptotic stability of the nonnegative (and nontrivial)
equilibrium $x^{::}$, where I is the subset of all indices at which $x_i^{::} = 0$. The proof
is complete.

The results of Theorem 2.1 and Theorem 2.2 still hold true if we insert in system (2.1)
interactions with continuous time delay, provided that the delay kernels have a parti
cular structure (A. Wörz-Busekros, 1978). We obtain the integrodifferential system:

$$\frac{dx_i}{dt} = x_i (e_i + \sum_{i \in N} a_{ij} x_j + \sum_{j \in N} \gamma_{ij} \int_{-\infty}^{t} F_{ij}(t-r) x_j(r) dr) + c_i \quad , \quad i \in N , \tag{2.28}$$

where $c_i > 0$ for all $i \in J \subseteq N$, $J \neq \emptyset$. We assume that the delay kernels are convex
combinations:

$$F_{ij}(t-r) = \sum_{k=1}^{P_{ij}} c_{ij}^{(k)} F_{ij}^{(k)}(t-r) \quad ; \quad c_{ij}^{(k)} \geq 0 \quad ; \quad \sum_{k=1}^{P_{ij}} c_{ij}^{(k)} = 1 \tag{2.29}$$

of the functions:

$$F_{ij}^{(k)}(t-r) = \frac{\alpha_{ij}^k}{(k-1)!} (t-r)^{k-1} \exp [-\alpha_{ij}(t-r)] \quad , \quad \alpha_{ij} \in R_+ \quad , \tag{2.30}$$

which are normalized to one:

$$\int_{0}^{+\infty} F_{ij}^{(k)}(s)ds = 1 \quad . \tag{2.31}$$

Let be $x_i(t) = x_i^{\ast}$, $i \in N$ for all $t \in (-\infty, +\infty)$ an equilibrium of (2.28).

By definition of the p-distinct supplementary variables

$$y_{ij}^{(k)}(t) := \int_{-\infty}^{t} F_{ij}^{(k)}(t-r)x_j(r)dr \quad , \quad k=1,\ldots,p_{ij} \quad ; \quad i,j \in N \tag{2.32}$$

that must attain at equilibrium the constant values $y_{ij}^{(k)\ast} = x_j^{\ast}$ for all k and i ,

system (2.38) is transformed into the system of O.D.E.

$$\begin{cases} \dfrac{dx_i}{dt} = x_i \left\{ e_i + \displaystyle\sum_{j \in N} a_{ij}x_j + \sum_{j \in N} \gamma_{ij} \sum_{k=1}^{p_{ij}} c_{ij}^{(k)} y_{ij}^{(k)} \right\} + c_i \quad , \quad i \in N \\[6pt] \dfrac{dy_{ij}^{(k)}}{dt} = \alpha_{ij} y_{ij}^{(k-1)} - \alpha_{ij} y_{ij}^{(k)} \quad , \quad k=1,\ldots,p_{ij} \quad ; \quad i,j \in N \end{cases} \tag{2.33}$$

where $y_{ij}^{(o)}(t) = x_j(t)$. Let $P = \{n+1,\ldots,n+p\}$ be the set of indices of the "p"

functions (2.32) and define:

$$y := \mathrm{col}\,(y_{ij}^{(k)}) \in R^{p} \quad , \quad x := \mathrm{col}\,(x_1,\ldots,x_n) \in R^{n'} \quad , \quad z := \mathrm{col}\,(x,y) \in R^{n+p}$$

$$\tag{2.34}$$

$$e^{(n)} := \mathrm{col}\,(e_1,\ldots,e_n) \in R^{n} \quad , \quad e := \mathrm{col}\,(e^{(n)},0) \in R^{n+p} \quad .$$

We obtain:

$$\begin{cases} \dfrac{dz_i}{dt} = z_i(e + Az)_i + c_i \quad , \quad i \in N \\[6pt] \dfrac{dz_i}{dt} = (Az)_i \quad , \quad i \in P \end{cases} \tag{2.35}$$

where A is an "$(n+p) \times (n+p)$" expanded community matrix which structure is obtained

by comparison between (2.33) and (2.35).

If $z^{\ast}$ is a nonnegative equilibrium of (2.35) and $I \subseteq N$ is the subset of all indi-

ces at which $z_i^{\ast} = 0$, by a simple generalization (adding $\dfrac{1}{2} \displaystyle\sum_{i \in P} w_i(z_i - z_i^{\ast})^2$, $w_i \in R_+$)

of the Liapunov function (1.10), we still obtain that the time derivative of the Lia-

punov function is given by (2.2) where now A is the expanded community matrix of

(2.35). We can use the same kind of homotopy techniques to prove:

Theorem 2.3. If in (2.35) $-A \in S_w$ and in $e^{(n)}$ there exists at least one index

$i \in N$ at which $e_i > 0$, then a nonnegative (and non-trivial) equilibrium $z^{\ast}$ of

(2.35) exists which is globally asymptotically stable in R_I^{n+p} .

Let us observe that if in (2.35) we set $c_i = 0$ for all i , Theorem 2.3 follows from Linear Complementary Problem (Solimano and Beretta, 1983).

We omitt the proof of Theorem 2.3 which is just an obvious extension of the proofs of Theorem 2.1 and Theorem 2.2 and can be found in Beretta and Solimano (1988).

3. Two different Volterra Patches Connected by Discrete Diffusion.

In a patch-type environment the habitat is divided into patches. Within each patch the spatial environment is sufficiently homogeneous that the dynamics can be described by Lotka-Volterra models and the individuals move between patches by discrete diffusion. (L.J.S. Allen, 1983).

We consider two different Volterra patches connected by discrete diffusion and described by the system of O.D.E.:

$$\frac{dx_i}{dt} = x_i(e_x + A_x x)_i + D_i(y_i - x_i) \quad , \quad i \in N$$

$$\frac{dy_i}{dt} = y_i(e_y + A_y y)_i + D_i(x_i - y_i) \quad , \quad i \in N \quad ,$$

(3.1)

where $N = \{1,...,n\}$, $D_i \geq 0$, $i \in N$ and there exists $k \in N : D_k > 0$. Furthermore we assume $e_x \neq e_y$, $A_x \neq A_y$. By the usual nomenclature and by definition of

$$\overline{e}_x := \text{col } (e_1^x - D_1,...,e_n^x - D_n) \in R^n \quad ; \quad \overline{e}_y := \text{col } (e_1^y - D_1,...,e_n^y - D_n) \in R^n \quad (3.2)$$

we may set system (3.1) in the form:

$$\frac{dx_i}{dt} = x_i(\overline{e}_x + A_x x)_i + D_i y_i \quad , \quad i \in N$$

$$\frac{dy_i}{dt} = y_i(\overline{e}_y + A_y y)_i + D_i x_i \quad , \quad i \in N$$

(3.3)

described by $z(t) = \text{col } (z(t),y(t)) \in R_{+o}^{2n}$. Let us observe that (3.3) belongs to the structure (1.9), where in vector (1.8) $c = 0$ and $B = \begin{pmatrix} 0 & D \\ D & 0 \end{pmatrix}$, with $D = \text{diag } (D_1,...,D_n)$.

Let be $N = 2n$ and let us define the homotopy function $H : R_{+o}^N \times T \rightarrow R^N$:

$$H_i(z,\xi) = x_i(\overline{e}_x + A_x x)_i + \xi D_i y_i + \xi(1-\xi)c_i \quad , \quad c_i > 0 \quad , \quad i \in N$$

(3.4)

$$H_i(z,\xi) = y_i(\overline{e}_y + A_y y)_i + \xi D_i x_i + \xi(1-\xi)d_i \quad , \quad d_i > 0 \quad , \quad i \in N$$

where $\xi \in T = [0,1]$. When $\xi = 0$ we have two decoupled Lotka-Volterra models and we regain the vector field of (3.3) when $\xi = 1$.

By the homotopy function (3.4) and following the same step procedure shown in the proof of Theorem 2.1, we can prove:

Theorem 3.1. If in (3.3)

(i) $-A_x$, $-A_y \in Z$ and have positive dominant diagonals (i.e. $-A_x$, $-A_y \in S_w$);

(ii) the diffusion coefficients D_i , $i \in N$ are sufficiently small that $\overline{e}_x$, $\overline{e}_y$ are positive vectors;

then system (3.3) has a positive equilibrium globally asymptotically stable in R_+^{2n} .

Proof. We recall that $-A \in Z$ means that $a_{ii} < 0$, $a_{ij} \geq 0$ for all $i,j \in N$.

By the homotopy function (3.4) we can prove that assumption (i) implies (H-1). Furthermore, assumption $-A_x$, $-A_y \in Z$ and assumption (ii) imply (H-2), i.e. $\dfrac{d\xi}{d\rho} > 0$ in $\widehat{\ker} H$. (H-1) and (H-2) imply the existence of a nonnegative equilibrium $z^{\ast} = \text{col } (x^{\ast}, y^{\ast})$ of (3.3) at which:

$$(\overline{e}_x + A_x x^{\ast})_i = \overline{e}_i^x + a_{ii}^x x_i^{\ast} + {\sum_{j \in N}}' a_{ij}^x x_j^{\ast} \leq 0 \quad , \quad i \in N$$

$$(\overline{e}_y + A_y y^{\ast})_i = \overline{e}_i^y + a_{ii}^y y_i^{\ast} + {\sum_{j \in N}}' a_{ij}^y y_j^{\ast} \leq 0 \quad , \quad i \in N \quad . \tag{3.6}$$

Since $-A_x$, $-A_y \in Z$ and $\overline{e}_x$, $\overline{e}_y$ are positive vectors, (3.6) clearly show that, if for some $i \in N$ $x_i^{\ast} = 0$ (or $y_i^{\ast} = 0$) , we get a contradiction because $(\overline{e}_x + A_x x^{\ast})_i > 0$ (or $(\overline{e}_y + A_y y^{\ast})_i > 0$) . Therefore, the equilibrium $z^{\ast}$ must have all positive components. By extending the Liapunov function (1.10) to both patches (1.11) becomes

$$\dot{V}\Big|_{R_+^{2n}} = (z - z^{\ast})^T W\widetilde{A}(z - z^{\ast}) - \sum_{i \in N} \frac{w_i^x D_i y_i (x_i - x_i^{\ast})^2}{x_i x_i^{\ast}} - \sum_{i \in N} \frac{w_i^y D_i x_i (y_i - y_i^{\ast})^2}{y_i y_i^{\ast}} \tag{3.7}$$

where $W = \text{diag } (w_1^x,\ldots,w_n^x;w_1^y,\ldots,w_n^y)$, w_i^x , $w_i^y \in R_+$, and according to (1.12)

$$\widetilde{A} = \begin{pmatrix} A_x & \dfrac{D_i}{x_i^{\ast}}\Big|i \in N \\ \dfrac{D_i}{y_i^{\ast}}\Big|i \in N & A_y \end{pmatrix} \tag{3.8}$$

The positivity of $z^{\ast}$ and the assumptions (i), (ii) are sufficient to prove (Beretta and Takeuchi, 1988) that $-\widetilde{A} \in S_w$, i.e., the global asymptotic stability of $z^{\ast}$ in

R_+^{2n} . The proof is complete.

Theorem 3.1 holds true even when in each patch we insert interactions with continuous time delay, providing that the delay kernels satisfy the assumptions (2.29)-(2.31) and that assumption (i) applies to the expanded community matrices of both the patches (Beretta and Takeuchi, 1988).

References

Allen, L.J.S.: Persistence and Extinction in Lotka-Volterra Reaction-Diffusion Equations. Math. Biosc., 65 , 1-12, (1983).

Beretta, E.; Capasso, V.: On the general structure of epidemic systems. Global asymptotic stability. Comp. & Maths. with Appls, 12A, 677-694, (1986).

Berman, A.; Plemmons, R.J.: Nonnegative Matrices in the Mathematical Sciences. Academic Press, New York, San Francisco, London, (1979).

Beretta, E.; Solimano, F.: A Generalization of Volterra Models with Continuous Time Delay in Population Dynamics: Boundedness and Global Asymptotic Stability. In press in SIAM J. Appl. Math., 48, n°3, June 1988.

Beretta, E.; Takeuchi, Y.: Global Asymptotic Stability of Lotka-Volterra Diffusion Models with Continuous Time Delay. In press in SIAM J. Appl. Math., 48, n°3, June 1988.

Garcia, C.B.; Zangwill, W.I.: Pathways to Solutions, Fixed Points and Equilibria, Prentice-Hall, Englewood Cliffs, N.J. (1981).

Solimano, F.; Beretta, E.: Existence of a Globally Asymptotically Stable Equilibrium in Volterra Models with Continuous Time Delay. J. Math. Biol. 18, 93-102, (1983).

Takeuchi, Y.; Adachi, N.: The Existence of Globally Stable Equilibria of Ecosystems of Generalized Volterra Type. J. Math. Biol., 10, 401-415, (1980).

Wörz-Busekros, A.: Global Stability in Ecological Systems with Continuous Time Delay. SIAM J. Appl. Math., 35, 123-134, (1978).

Acta Applicandae Mathematicae **14** (1989), 49–57.
© 1989 *by IIASA.*

Cooperative Systems Theory and Global Stability of Diffusion Models

Y. Takeuchi

Department of Applied Mathematics

Faculty of Engineering

Shizuoka University

Hamamatsu 432, Japan

AMS Subject Classification (1980): 92A17, 34D20
Key words: global stability, cooperative systems, diffusion

1. Introduction

Many authors consider the effect of spatial factors, such as diffusion or migration among patches, in population dynamics. We suppose that the system is composed of several patches connected by diffusion and occupied by a single species. Furthermore, the species is supposed to be able to survive in all the patches at a positive globally stable equilibrium point if the patches are isolated, or if the diffusion among patches is neglected and the species is confined to each patch. The problem considered in this paper is whether the equilibrium point, the value of which can be changed according to the strength of diffusion, continues to be positive and globally stable, if we increase the rates of diffusion.

Allen [1] proved by applying comparison techniques that the model of such a single species diffusion system remains weakly persistent if the strength of diffusion is *small* enough. The homotopy function technique was successfully applied by Beretta and Takeuchi [2,3] to show that *small diffusion* cannot change the global stability of the model. On the other hand, Hastings [6] proved that the positive equilibrium point of the model, if it exists, is locally stable for sufficiently *large diffusion*. These results are valid for general multiple patch models. For the model restricted to a two patch system, Freedman, Rai and Waltman [4] showed that there exists a positive equilibrium for *any diffusion* rate and that it is globally stable if it is unique.

These known results may suggest that diffusion cannot change the global stability of the model and the purpose of this paper is to show that the model continues to be globally stable for *any diffusion* rate.

2. The Model and Cooperative Systems Theory

The model considered in this paper is described by the following system of ordinary differential equations:

$$\dot{x}_1 = \dot{x}_i g_i(x_i) + \sum_{j=1}^{n} D_{ij}(x_j - x_i) \qquad x_i(0) > 0 \qquad (2.1)$$

where $i=1,\ldots,n$, n is the number of the patches in the system, x_i represents the population density of the species in the i-th patch and $g_i(x_i)$ represents the specific growth rate of the population in the i-th patch. Since the specific growth rate may depend on each patch environment, the function $g_i(x_i)$ is supposed to be different in each patch. The second term of the right hand side in (2.1) describes the diffusion effect between patches: D_{ij} is a nonnegative diffusion coefficient for the species from j-th patch to i-th patch ($i \neq j$) and $D_{ii} = 0$ ($i=1,\ldots,n$). It is supposed in model (2.1) that the net exchange from the j-th patch to the i-th patch is proportional to the difference $x_j - x_i$ of population densities in each patch. This is the usual assumption (see, e.g. [1,2,3,6]).

Furthermore, we suppose the following:

(H1) All solutions of the initial value problem (2.1) exist, are unique and are continuable for all positive time.

(H2) $g_i(0) > 0$, $g_i'(x_i) < 0$ and $x_i g_i(x_i) < 0$ as $x_i \to +\infty$, $i=1,..,n$.

(H3) The matrix $D = (d_{ij})$ is irreducible, where for $i, j=1,\ldots,n$,

$$d_{ij} = \begin{cases} D_{ij} & \text{for } j \neq i \\ -\sum_{j=1}^{n} D_{ij} & \text{for } j=i \end{cases}$$

The hypotheses (H1) and (H2) are standard in single species models [4]. We need (H3) to ensure the boundedness of all solutions to (2.1) (see Lemma 1 in the next section.) The irreducibility of matrix D implies that every patch in (2.1) is connected by diffusion and that the only possible equilibrium points are the origin and the positive one.

Note that model (2.1) without diffusion ($D_{ij} = 0$ for all $i,j = 1,\ldots,n$) has a positive globally stable equilibrium point $(K_1,\ldots,K_n)$ by (H2), where K_i is a positive scalar satisfying $g_i(K_i) = 0$, $i = 1,\ldots,n$. The problem considered in this paper is whether the equilibrium continues to be positive and globally stable for any $D_{ij} \geq 0$ ($i,j = 1,\ldots,n$) or not.

We use the following notations in this paper: the right hand side of model (2.1) is denoted by $F_i(x)$, that is,

$$F_i(x) = x_i g_i(x_i) + \sum_{j=1}^{n} D_{ij}(x_j - x_i) , \quad i=1,\ldots,n \tag{2.2}$$

and two important subsets of $R_+^n = \{x \mid x_i > 0, i=1,\ldots,n\}$ are defined by $X^+ = \{x \in R_+^n \mid F_i(x) > 0, i=1,\ldots,n\}$ and $X^- = \{x \in R_+^n \mid F_i(x) < 0, i=1,\ldots,n\}$. If model (2.1) has a positive equilibrium point x^*, we define the sets $\bar{X}^* = \{x \mid x > x^*\}$, $\underline{X}^* = \{x \in R_+^n \mid x < x^*\}$ and $H_{x^*}^i = \{x \in R_+^n \mid x \geq x^*, x_i = x_i^*\}$. Note that the boundary of $\bar{X}^*$ lies in the n hyperplanes $H_{x^*}^i, i=1,\ldots,n$.

The most important property of (2.1) is that it is a so called quasimonotone [5] or cooperative [7, 10] system, since

$$\partial F_i(x)/\partial x_j = D_{ij} \geq 0 \quad \text{for any } i \neq j . \tag{2.3}$$

For single species diffusion models such as (2.1) this property has so far been neglected. But it is important since such a quasimonotone or cooperative model can have only a relatively simple dynamical behavior (see [5,7]). For example, cooperative systems cannot have attracting cycles [5,7]. Furthermore, if the cooperative system

$$\dot{x}_i = H_i(x), \; H_i(0) = 0, \; i = 1,\ldots,n \tag{2.4}$$

has the following three properties:

(i) $DH(x)$ is irreducible for any $x \geq 0$, (2.5)

(ii) $DH(x) \leq DH(y)$ for any $x \geq y \geq 0$ and

(iii)   all solutions are bounded,

then either the origin is globally stable or else there exists a unique positive equilibrium point and all the trajectories in $R_+^n \backslash \{0\}$ tend to it [7]. Here $DH(x)$ is the Jacobian of $H(x)$. For the cooperative system

$$\dot{x}_i = x_i h_i(x) , \tag{2.6}$$

Smith [10] assumed a similar condition

$$\text{if } x \geq y \geq 0, \text{ then } Dh(x) \leq Dh(y) . \tag{2.7}$$

Conditions (2.5) or (2.7) imply that cooperative effects diminish with increasing population density [7, 10].

Unfortunately, model (2.1) does not satisfy (2.5) or (2.7) under our assumptions (H1), (H2) and (H3). If we suppose further

$$g_i^{''}(x_i) \leq 0 \text{ for all } x_i \geq 0, \ i = 1,\ldots,n \tag{2.8}$$

and if there exists a positive equilibrium point, then it is a globally stable equilibrium point of (2.1) by Hirsch's theorem [7]. In general, condition (2.8) is not satisfied in population dynamics. For example, consider the following cooperative hunting model:

$$\dot{x}_i = x_i(K_i - x_i)^3, \quad K_i > 0, \quad i = 1,\ldots,n , \tag{2.9}$$

which does not satisfy (2.8). But model (2.9) obviously has a positive and globally stable equilibrium point $(K_1,\ldots,K_n)$. The main theorem in the next section shows that the positive and globally stable equilibrium point of model (2.9) with diffusion term continues to exist for any diffusion rate. To prove the global stability of (2.1) without assumptions (2.8), the monotone property of flows of cooperative systems [5,7,8,10] still plays an important role.

3. Main Results

In this section we assume (H1), (H2) and (H3) for model (2.1) and prove the existence of a positive and globally stable equilibrium point of (2.1) for any diffusion rate $D_{ij} \geq 0 \ (i,j = 1,\ldots,n)$.

First we note that for each $i = 1,\ldots,n$, $F_i(x) \geq 0$ if $x_i = 0$ and all $x_j \geq 0 \ (j \neq i)$, which ensures the positive invariance of R_+^n. The boundedness of the solutions to (2.1) is shown by the following lemma.

Lemma 1. All the solutions to (2.1) are bounded.

Proof. It is trivial that every row sum of matrix D in (H3) vanishes and that all the off-diagonal elements of D are nonnegative. These properties and the irreducibility of matrix D ensure the existence of a positive vector c such that $c^T D = 0$ [9]. Hence the weighted sum of $\dot{x}_i$ becomes

$$\sum_{i=1}^{n} c_i \dot{x}_i = \sum_{i=1}^{n} c_i x_i g_i(x_i) < 0 \text{ for sufficiently large } x_i, \ i = 1,\ldots,n . \tag{3.1}$$

Since $c_i > 0$, $i = 1,\ldots,n$, (3.1) shows the boundedness of the solutions.

Lemma 2. For any point $\underline{x} = (\epsilon, \ldots, \epsilon) > 0$ where ϵ is a positive scalar, there exists a scalar $\underline{a} > 0$ such that

$$F_i(a\underline{x}) > 0 \text{ for any } 0 < a < \underline{a}, \quad i = 1, \ldots, n . \tag{3.2}$$

Proof. For any scalar $a > 0$,

$$F_i(a\underline{x}) = a\epsilon g_i(a\epsilon) \tag{3.3}$$

by the definition of $\underline{x}$. At $a=0$,

$$\partial F_i(a\underline{x})/\partial a = \epsilon g_i(0) > 0 \tag{3.4}$$

$$F_i(0) = 0, \quad i = 1, \ldots, n , $$

which ensures the existence of an $\underline{a}$ satisfying (3.2).

It is obvious that the set $\{x \,|\, x \geq u\}$ is positively invariant if u is any point belonging to the set $\{x = a\underline{x} \,|\, 0 < a < \underline{a}\}$. In fact, for any $y \in \{x \,|\, x \geq u\}$ and $y_i = u_i$,

$$F_i(y) \geq F_i(u) > 0 \tag{3.5}$$

where the first inequality in (3.5) holds by (2.3) and the second by (3.2). Therefore, Lemmas 1 and 2 imply the following theorem:

Theorem 3. The model (3.1) is persistent for any $D_{ij} \geq 0$, that is, any solution $x(t)$ satisfies

$$\lim_{t \to \infty} \inf x_i(t) > 0, \quad i = 1, \ldots, n \tag{3.6}$$

and furthermore there exists a positive equilibrium point.

Theorem 3 implies that the species can survive in all patches under any diffusion rate and that (2.1) has at least one positive equilibrium point. Next we prove that it is unique in R_+^n.

Lemma 4. A positive equilibrium point of (2.1) is unique.

Proof. Let us denote a positive equilibrium of (2.1) by $\hat{x}$ and consider the function $F_i(a\hat{x})$ for a positive scalar a. Then

$$F_i(a\hat{x}) = a\hat{x}_i g_i(a\hat{x}_i) + \sum_{j=1}^{n} D_{ij}a(\hat{x}_j - \hat{x}_i) \tag{3.7}$$

$$= a\hat{x}_i[g_i(a\hat{x}_i) - g_i(\hat{x}_i)], \quad i = 1,\ldots,n .$$

Here the last equality is obtained because $\hat{x}$ is an equilibrium point of (2.1). By (H2),

$$F_i(a\hat{x}) \begin{cases} > 0 & \text{for } 0 < a < 1 \\ < 0 & \text{for } a > 1 . \end{cases} \tag{3.8}$$

Since the vector $a\hat{x}(a > 0)$ is a half-line starting at the origin, pointing into the positive cone R^n_+ and passing through $\hat{x}$ at $a = 1$, (3.8) implies for all $i = 1,\ldots,n$ that $F_i(x) > 0$ on $x = a\hat{x}$ if $0 < a < 1$ and $F_i(x) < 0$ on $x = a\hat{x}$ if $a > 1$.

Suppose that there exists another equilibrium point $\tilde{x} > 0$, $\tilde{x} \neq \hat{x}$ and that

$$\hat{x}_j < \tilde{x}_j \quad \text{for some } j \in \{1,\ldots,n\} . \tag{3.9}$$

Since $F_i(x)$ is nondecreasing in $x_k(k \neq i)$, for any $x \in H^i_{\tilde{x}}$,

$$0 = F_i(\tilde{x}) \leq F_i(x_1,..,x_{i-1}, \tilde{x}_i, x_{i+1},\ldots,x_n) = F_i(x) , \quad i = 1,\ldots,n . \tag{3.10}$$

On the other hand, the vector $a\hat{x}$ for sufficiently large $a > 1$ must intersect one of the $H^i_{\tilde{x}}$ (or intersections of some $H^i_{\tilde{x}}$) by (3.9). But at the point x of the intersection, $F_i(x) < 0$ by (3.8), which contradicts (3.10). When $\tilde{x}_j \leq \hat{x}_j$ for any $i = 1,\ldots,n$, we consider hyperplanes $H^i_{\hat{x}}$ and the vector $a\tilde{x}$. Again we get a contradiction. This completes the proof of Lemma 4.

Hereafter we denote a unique positive equilibrium as x^*. From (3.8) ($\hat{x} = x^*$), both sets X^+ and X^- are nonempty. Note that X^+ and X^- can be written as $X^+ = X^+_1 \cup X^+_2$ and $X^- = X^-_1 \cup X^-_2$ where X^+_i and $X^-_i(i = 1,2)$ are open and disjoint, $X^+_1 \supset \{x \mid x = ax^*, 0 < a < 1\}$ and $X^-_1 \supset \{x \mid x = ax^*, a > 1\}$. X^+_2 and X^-_2 may be empty. We denote X^+_1 as X^+, and X^-_1 as X^- for simplicity. The next lemma shows that both X^+ and X^- are attracting regions of x^*, or that x^* is globally stable with respect to X^+ and X^-.

Lemma 5. There exist nonempty sets $X^+ \subseteq X^*$ and $X^- \subseteq \bar{X}^*$. Any solution of (2.1) initiating in X^+ or X^- converges to x^* as $t \to +\infty$.

Proof. By (3.8), X^+ and X^- are not empty. By (2.3) and (3.10), $F_i(x) \geq 0$ on $H_{x^*}^i$ for all $i = 1, \ldots, n$ which shows that $X^- \subseteq \bar{X}^*$. Similarly we can prove that $X^+ \subseteq \underline{X}^*$.

Now let us prove that any solution initiating in X^+ cannot leave X^+ in finite time. Consider any solution $x(t)$ starting in X^+ and suppose that $x(t)$ reaches a boundary point of X^+ for the first finite time $0 < T < +\infty$. Furthermore for simplicity let us assume that

$$F_i(x(T)) \begin{cases} = 0 & i \in M = \{1, \ldots, m\}(\neq \phi) \\ > 0 & i \in N - M = \{m+1, \ldots, n\}(\neq \phi) \end{cases} \tag{3.11}$$

For any $i \in M$, by (2.2) and (3.11),

$$dF_i(x(t))/dt \Big|_{t=T} = \sum_{j \in N-M} D_{ij} F_j(x(T)) . \tag{3.12}$$

By (H3), there exists for each $i \in M$ at least one $j \in N - M$ such that $D_{ij} > 0$. Therefore, $dF_i/dt > 0$ at $t = T$ for all $i \in M$, which contradicts (3.11). Hence any solution initiating in X^+ cannot reach a boundary point of X^+ at finite time.

Next let us define a continuously differentiable function $V(x)$:

$$V(x) = \sum_{i=1}^n (x_i^* - x_i) \geq 0 \text{ for any } x \in X^+ . \tag{3.13}$$

Here the last inequality holds since $X^+ \subseteq \underline{X}^*$. The equality holds only for $x = x^*$. Then the time derivative of $V(x)$ along a solution of (2.1) satisfies the following relationship:

$$dV(x)/dt \Big|_{(2.1)} = - \sum_{i=1}^n F_i(x) < 0 \text{ in } X^+ \tag{3.14}$$

where the last inequality holds for any finite time and for any solution $x(t)$ starting in X^+. Furthermore the equality

$$\sum_{i=1}^n F_i(x) = 0 \text{ in } X^+ \cup \partial X^+ \backslash \{0\} \tag{3.15}$$

holds only for $x = x^*$ since x^* is the unique equilibrium point in $X^+ \cup \partial X^+ \backslash \{0\}$. This shows that x^* is globally stable with resspect to X^+. Similarly we can prove that x^* is also globally stable with respect to X^-. This completes the proof of Lemma 5.

From the lemmas above, finally, we can prove the gobal stability of a positive equilibrium point x^* with respect to R_+^n.

Theorem 6. The equilibrium x^* is globally stable with respect to R_+^n.

Proof. It is trivial that we can choose two points $\alpha \in X^+$ and $\beta \in X^-$ for any $x^0 \in R_+^n$ such that

$$\alpha < x^0 < \beta \qquad (3.16)$$

by (3.8) and the definitions of X^+ and X^-. Let us denote by $x^\alpha(t)$, $x(t)$ and $x^\beta(t)$ the solutions initiating at $x = \alpha$, x^0 and β, respectively. Then by Kamke's theorem for cooperative systems [8], they satisfy

$$x^\alpha(t) \leq x(t) \leq x^\beta(t) \quad \text{for any } t \geq 0 . \qquad (3.17)$$

Furthermore, from Lemma 5, $x^\alpha(t) \to x^*$ and $x^\beta(t) \to x^*$ as $t \to +\infty$. Therefore, any solution $x(t)$ initiating in R_+^n must converge to x^* as $t \to +\infty$ by (3.17). The stability of x^* can be shown easily from the global stability of x^* with respect to X^+ and X^-, and from (3.17).

4. Conclusion

We have proved that the single species diffusion model has always a positive and globally stable equilibrium point for any diffusion rate. Since the model has a positive and globally stable equilibrium point when all the patches in the system are isolated, the result obtained in this paper shows that no diffusion rate can change the global stability of the model. In order to show the global stability, the important properties of the model are relationship (3.8) and the monotonicity of flows in cooperative systems (i.e., (3.16) and (3.17)). When we consider two patches connected by diffusion and assume that each patch has only symbiotic interactions, Kamke's theorem still holds but a property similar to (3.8) is only true if the diffusion between the two patches is weak enough. Therefore, we can prove the global stability of such a model for weak diffusion.

In this paper, we assumed the simplest linear diffusion form. If we adopt nonlinear diffusion forms, property (3.8) becomes more difficult to check. We need some conditions for ensuring weak nonlinearity or we must assume the uniqueness of a positive equilibrium point.

Acknowledgement. Financial support from the International Institute for Applied Systems Analysis during my stay in Vienna is gratefully acknowledged. I have greatly benefited from the discussions with K. Sigmund, J. Hofbauer, G. Kirlinger and E. Beretta in Vienna.

References

[1]   Allen, L.J.S. (1987), Persistence, extinction, and critical patch number for island populations, J. Math. Biol., 24, 617-625.

[2]   Beretta, E. and Takeuchi, Y. (1987), Global stability of single-species diffusion models with continuous time delays, Bull. Math. Biol., 49, No. 4, 431-448.

[3]   Beretta, E. and Takeuchi, Y. (1988), Global asymptotic stability of Lotka-Volterra diffusion models with continuous time delay, SIAM J. Appl. Math., 48, No. 3, 627-651

[4]   Freedman, H.I., Rai, B., and Waltman, P. (1986), Mathematical models of population interactions with dispersal II: Differential survival in a change of habitat, J. Math. Anal. Appl., 115, 140-154.

[5]   Hadeler, K.P. and Glas, D. (1983), Quasimonotone systems and convergence to equilibrium in a population genetic model, J. Math. Anal. Appl., 95, 297-303.

[6]   Hastings, A. (1982), Dynamics of a single species in a spatially varying environment: The stabilizing role of higher dispersal rates, J. Math. Biol., 16, 49-55.

[7]   Hirsch, M.W. (1984), The dynamical systems approach to differential equations, Bull. A.M.S., 11, No. 1, 1-634.

[8]   Kamke, E. (1932), Zur Theorie der Systeme gewöhnlicher Differentialgleichungen II, Acta Math., 58, 57-85.

[9]   Nikaido, H. (1968), Convex structure and economic theory, Academic Press, New York - London.

[10] Smith, H.L. (1986), On the asymptotic behavior of a class of deterministic models of cooperating species, SIAM J. Appl. Math., 46, 368-375.

Acta Applicandae Mathematicae **14** (1989), 59–74.
© 1989 *by IIASA*.

The n-Person War of Attrition

John Haigh

Mathematics Division
University of Sussex
Falmer, Brighton BN1 9QH, U.K.

Chris Cannings

Department of Probability & Statistics,
The University
Sheffield S3 7RH, U.K.

AMS Subject Classification (1980): 92A12
Key words: evolutionarily stability, war of attrition, strategies

1. Introduction

The War of Attrition (WA) was one of the earliest examples studied in the use of the theory of games to understand animal behavior (see Maynard Smith (1974)). The setup is that two contestants compete for a prize worth $V(V > 0)$, and the one who is prepared to wait longer collects the prize; both contestants incur a cost equal to the length of time taken to resolve the contest. Symbolically, if $E(x,y)$ denotes the amount gained by a contestant prepared to wait time x when the opponent is prepared to wait y,

$$E(x,y) = \begin{cases} V-y & \text{if } x > y \\ -x & \text{if } x < y \end{cases} \tag{1}$$

with

$$E(x,x) = \frac{V}{2} - x \qquad x\in[0,\infty)$$
$$y\in[0,\infty) \ .$$

Such a game has precisely one evolutionarily stable strategy or ESS (Bishop and Cannings (1976)), i.e. a strategy such that if played by a population, no mutant using another strategy can invade. This ESS is to wait for a time x drawn at random from the exponential distribution with mean V, i.e. density $\frac{1}{V} \exp(-x/V)$ $(x > 0)$.

Several modifications and generalisations of the basic model (1) have been investigated (e.g. Bishop and Cannings (1978), (1986), Haigh and Rose (1980), Maynard Smith and Parker (1976)). Here we analyze several different models for the generalisation of (1) to n-person conflicts. The common feature of the models is that the contestants pay a cost, measured as the length of time they participate in the contest, and are prepared to do so because of their hope of receiving a reward. An ESS for a n-person WA (if one exists) will be a probability distribution G, that governs the length of time a contestant is prepared to wait, with the property that, if all the population are using G, then no invader who uses a different distribution H can establish himself from a small base, i.e. a small fraction ϵ of the population.

The various models described below hopefully reflect some of the situations which might arise in nature. There are many situations in which a number of individuals $(n > 2)$ are simultaneously in conflict; a pack with a kill must establish the order in which individuals may feed (though a prior "pecking order" may exist); territorial animals will need to establish how a set of potential territories are allocated; male bees may compete for the right to mate with the queen. These examples suffice to indicate that there are occasions in which n individuals compete, and that there may be only a single reward available, or a number of rewards, and provide the motivation for the choice of models below.

The only prior work of which we are aware is some discussion by Palm (1984) of the general notion of an ESS with n contestants, but no specific models are considered.

2. The Models

Model A. With n players competing for one reward, value V, each player independently selects a time he is prepared to wait, in the hope of outlasting the $(n-1)$ opponents. Once chosen, this time is fixed: as some players drop out, those remaining are not allowed to alter their initial "bids".

Here player i will select a random time X_i drawn from some distribution G that depends only on V and n. The reward to player 1 can be written

$$E(X_1;X_2,\ldots,X_n) = \begin{cases} V-W_1 & \text{if } X_1 > W_1 \\ -X_1 & \text{if } X_1 < W_1 \end{cases} \tag{2}$$

where $W_1 = Max(X_2,X_3,\ldots,X_n)$, and the reward to the other players follow in a symmetric manner. We can safely ignore consideration of the case $X_1 = W_1$ since this has probability zero of arising when the distribution of $\{X_1\}$ is continuous; and this is the case in this class of contests, because no ESS can have an atom of probability anywhere (Bishop and Cannings (1978)).

Theorem 1. Model A has a unique ESS, when each player chooses a value, independently, from the distribution functions

$$G(x) = (1 - \exp(-x/V))^{\frac{1}{n-1}} \quad (x \geq 0). \tag{3}$$

Proof. From the point of view of the user of X_1, he is in a 2-person WA with an opponent who uses strategy W_1. For such a game, a value x is in the support of an ESS only if $E(x, W_1)$ is independent of x, and so, as in the 2-person WA, the only candidate for W_1 is the exponential distribution with mean V, i.e. $P(W_1 \leq t) = 1 - \exp(-t/V)$. Since $P(W_1 \leq t) = P(X_1 \leq t)^{n-1}$, equation (3) describes the only candidate G for an ESS.

Consider a population, a fraction $1-\epsilon$ of which use G and fraction ϵ use some other strategy described by the distribution function H. Here, the mean returns to different types from n-player WA's, when contestants drawn at random from the population, are:

$$E(\text{Return to } G\text{-player}) = (1-\epsilon)^{n-1}E(G; G^{(n-1)}) \tag{4A}$$

$$+ \sum_{r=1}^{n-1} \binom{n-1}{r} \epsilon^r (1-\epsilon)^{n-r-1} E(G; H^{(r)}, G^{(n-1-r)})$$

and

$$E(\text{Return to } H\text{-player}) = (1-\epsilon)^{n-1}E(H; G^{(n-1)}) \tag{4B}$$

$$+ \sum_{r=1}^{n-1} \binom{n-1}{r} \epsilon^r (1-\epsilon)^{n-r-1} E(H; H^{(r)}, G^{(n-1-r)})$$

where $E(A; B^{(r)}, C^{(s)})$ denotes the mean return to an A-player when the opponents are r B-players and s C-players.

Since $E(x; G^{(n-1)})$ is constant for all x, the first terms on the right in (4A), (4B) are equal. Thus for a small ϵ, $E(\text{Return to } G\text{-player})$ exceeds $E(\text{Return to } H\text{-player})$ if, and only if, $I > 0$, where

$$I = E(G; H^{(1)}, G^{(n-2)}) - E(H; H^{(1)}, G^{(n-2)}) . \tag{5}$$

This has previously been demonstrated by Palm (1984) and is included here for completeness.

Suppose $T(x)$ is the mean return from using pure strategy x in an n-person WA when the opponents consist of one H-player and $(n-2)$ G-players. Let Y and $X_3, X_4, \ldots, X_n$ be random variables representing the strategies used by these opponents and write $W = Max(Y, X_3, \ldots, X_n)$. Thus

$$T(x) = \int_0^x (V-w)\varphi(w)\,dw - x \int_x^\infty \varphi(w)\,dw \qquad (6)$$

where $\varphi(w)$ is the density of W. Now $P(W \le w) = H(w)G(w)^{n-2}$, so one integration by parts on the first term of (6) leads to

$$T(x) = VH(x)G(x)^{n-2} - x + \int_0^x H(w)G(w)^{n-2}\,dw .$$

But $I = \int_0^\infty T(x)[g(x) - h(x)]\,dx$, so

$$I = \int_0^\infty \{[VH(x)G(x)^{n-2} - x](g(x) - h(x)) - [G(x) - H(x)]H(x)G(x)^{n-2}\}\,dx \qquad (7)$$

$$= \int_0^\infty f(x, H, h)\,dx$$

(say), since G, g are known functions. To show that $I > 0$ unless $H = G$ we will show, using the calculus of variations (see, e.g. Gelfand and Fomin (1963)) that I is minimised when $H = G$. The Euler equation

$$\frac{d}{dx}\left(\frac{\partial f}{\partial h}\right) - \frac{\partial f}{\partial H} = 0$$

is

$$\frac{d}{dx}[-VHG^{n-2} + x] - VG^{n-2}(g-h) + [G-H]G^{n-2} - HG^{n-2} = 0 .$$

[Here, and later, we contract $G(x)$ to G, etc., where this is not ambiguous.] Hence,

$$-(n-2)VHgG^{n-3} + 1 - VgG^{n-2} + G^{n-1} - 2HG^{n-2} = 0 . \qquad (8)$$

Also, from (3), $G^{n-1} = (1-\exp(-x/V))$, so $(n-1)VgG = \exp(-x/V)$. Hence (8) can be written

$$H[2-2\exp(-x/V) + \frac{n-2}{n-1}\exp(-x/v)] = G[2-\exp(-x/V) - \frac{1}{n-1}\exp(-x/V)]$$

whose only solution is $H(x) = G(x)$.

To show this is a minimum, it is enough to show that

$$J(\alpha) \equiv \int_0^\infty (P\alpha'^2 + Q\alpha^2)\,dx \geq 0 \text{ for all functions } \alpha \,,$$

where

$$P = 1/2\,\frac{\partial^2 f}{\partial h^2} \text{ and } Q = 1/2\,(\frac{\partial^2 f}{\partial H^2} - \frac{d}{dx}\,(\frac{\partial^2 f}{\partial h \partial H}))\,.$$

Here

$$P = 0,\ \ Q = 1/2\,[2G^{n-2} - \frac{d}{dx}\,(-VG^{n-2})] = 1/2\,[2G^{n-2} + (n-2)\,VgG^{n-3}]\,.$$

Hence the condition on J is clearly satisfied and, since I only has one local extremal, $I \geq 0$ for all choices H.

One feature of the two-person WA is that the total mean gain to the population from the contest is zero, because the value V of the prize is exactly offset by the mean losses, each player losing, on average, $V/2$, the mean duration of the game. Our next result is that this also holds for the current Model.

Theorem 2. If, in Model A, all members of the population use the ESS, then the sum of the times waited until the game ends has mean V.

Proof. This follows immediately since the support of the ESS is $[0,\infty)$, hence $E(x; G^{(n-1)})$ is a constant all $x \geq 0$, and this must be zero since playing $x = 0$ clearly has payoff zero. Thus the expected payoff for each individual is zero in an "ESS-population", so total time used is V, exactly cancelling the reward value.

One reaction to this result may be to feel that there is no incentive to a contestant to enter this contest, as the mean gain to each contestant is zero. But this reaction would be misplaced: the incentive to enter the contest is the prospect of gaining reward V, and, in order to do so, contestants are prepared to invest some time.

It is appropriate to point out that if all the individuals are playing the ESS, then their plays are $X_1, X_2, \ldots, X_n$ which are independent random variables with distribution function G. Supposing that $X_{(1)} < X_{(2)} < \cdots < X_{(n)}$ are the corresponding order statistics then the total time expended is

$$T = X_{(1)} + X_{(2)} + \cdots + X_{(n-1)} + X_{(n-1)}$$

(since the contest ends when the $(n-1)^{\text{th}}$ contestant quits). It can be demonstrated directly that this T has expected value V for G as given in Theorem 1, but such a proof will essentially mirror the equation $\int_0^\infty g E(X ; G^{(n-1)}) \, dx = 0$, and so does not add greatly to our understanding. On the other hand the expression for T allows us to express the length of a contest as

$$E(X_{(n-1)}) = T - n E(X) + E(X_{(n)})$$

which, since $T = V$, can then be expressed as

$$E(X_{(n-1)}) = V\{1 - n\psi\left(\frac{n}{n-1}\right) + \psi\left(\frac{2n-1}{n-1}\right) - (n-1)\gamma\}$$

where $\psi(Z)$ is the Psi function, and γ is Euler's constant (see for example Abramowitz and Segun (1965)).

This can be evaluated for particular values of n, giving for $n = 2$ the value $V/2$, and for $n = 3$ we have $V\{4ln2 - 7/3\} \approx 0.44\,V$. As expected the average length of contest must decrease the greater the number of contestants.

Model B. This differs from Model A only in that, each time one player drops out, the other players notice this and are allowed to reappraise and alter their "bids". Thus, when n players $(n \geq 3)$ compete, player i selects a time $X_i^{(n)}$ he is prepared to wait (his "bid"). At time Min $(X_1^{(n)}, X_2^{(n)}, \ldots, X_n^{(n)})$, one player drops out, and now player i (if he remains in the game) selects a bid $X_i^{(n-1)}$ that he is *further* prepared to wait.

It will be seen that Model B is a special case of Model C, but it is useful to consider Model B separately. Any strategy for the game is a list of the strategies a player would use in a r-person WA, for $r = 2, 3, \cdots$. When $r = 2$, we have the familiar 2-person WA, with its exponential ESS and the property that the mean reward to the two participants is zero. Since a player can only gain any reward if he survives to participate in the 2-person WA, where he can expect to be faced by an opponent who uses the exponential ESS, any strategy $X^{(r)}$ (for $r \geq 3$) which leads to the expenditure of any time at all before the game reduces to two players would have a strictly negative mean gain, and thus be inferior to the simple strategy of quitting immediately when $r \geq 3$.

Let $(E ; Q)$ denote the strategy: "use the exponential distribution with mean V when $r = 2$, and quit immediately when $r \geq 3$" . Similarly, let $(F ; R)$ be a strategy, where F is some strategy for $r = 2$ *other than* this exponential, and R is some strategy for $r = 3$. Clearly, since $(F ; R)$ is inferior to $(E ; R)$ no such $(F ; R)$ can be an ESS. Thus the only candidates for an ESS are strategies of the form $(E ; R)$ (including, in particular, $(E ; Q)$).

In any population containing both $(E;Q)$ and $(E;R)$ with R different from "Quit immediately", users of $(E;Q)$ are at a disadvantage. But, when $(E;R)$ is rare, so that we can ignore the possibility of three or more such players entering the same contest, it is not at a strict disadvantage, so there is no selective pressure to eliminate it from the population. We must examine carefully what we would expect to happen in a n-player contest $(n \geq 3)$ in which all players (or all but one or two players) use $(E;Q)$.

One possibility is that the decision to Quit is executed immediately, all who use $(E;Q)$ leaving the scene of battle as soon as the contest begins. If this occurred, and two $(E;R)$ players were left, they would contest an ordinary two-person WA; if one $(E;R)$ player were left, he would collect V without a contest; and if no-one were left, the prize would go unclaimed. Clearly, $(E;R)$ is at an advantage, especially when rare, and so would be sustained (at a low level) in the population.

But this scenario is inherently implausible; to imagine that all $(n-1)(E;Q)$ players who quit would vacate the scene instantly, and allow one $(E;R)$ player to enjoy V without opposition, or allow the prize to go unclaimed, does not carry conviction. An alternative approach is to say that, if k players take the decision to Quit, random chance selects one of them to be the first to carry out this decision (instantaneously), and then, immediately, there is one fewer contestant and all remaining players reappraise their bids. With this interpretation, any mixture of $(E;Q)$ and $(E;R)$ players (at most two of the latter) would be by a series of instantaneous decisions, reduced to two contestants who would play an ordinary WA. Here $(E;R)$ carries no advantage at any frequency level, and is disadvantaged unless rare.

We summarise our results as a theorem.

Theorem 9. In Model B, there is no ESS. The strategy $(E;Q)$, if played by the population, cannot be displaced. If we interpret the simultaneous decision of several players to Quit immediately as being implemented by one of them (chosen at random) immediately, to be followed by a strategy reappraised by those remaining, then the other strategy is at a disadvantage at any frequency level and will be eliminated by random drift.

If, alternatively, we suppose that the simultaneous decision by several players to Quit immediately is simultaneously and immediately implemented by them all, then strategies of the form $(E;R)$, $R \neq Q$, are advantaged when rare (but cannot reach even moderate frequency).

motivated to look at an n- person WA in which rewards $V_n, V_{(n-1)}, \ldots, V_1$ become successively available. As in Model B, we suppose that players are allowed to reappraise their strategy as the number of contestants decreases. The game then proceeds as follows: all n players select a time they are prepared to wait; the first person whose waiting time expires collects reward V_n and leaves; the remaining $n-1$ players now play a similar game until, eventually, when the last but one player drops out with reward V_2, the sole remaining player immediately takes V_1. A strategy for a player is a description of how long you are initially prepared to wait for any collection of rewards $(V_n, V_{n-1}, \ldots, V_1)$.

Theorem 4. Suppose, in Model C, $V_n < V_{n-1} < \cdots < V_1$. Then there is a unique ESS: wait for time X, where X is an exponential random variable with mean $(n-1)(V_{n-1} - V_n)$. Further, the value of this game to each player, $(= E$ (Player's reward)) is V_n.

Proof. When $n = 2$, we have a two-person WA with rewards V_2, V_1 which has the same ESS as if the rewards are 0 and $V_1 - V_2$, i.e. exponential with mean $V_1 - V_2$. The expected reward for the latter game is 0, hence for the former is V_2 to each. This proves the result for $n = 2$.

Suppose inductively the result holds for $n-1$ rewards, and consider the case of n rewards. If each player uses the given exponential strategy, the length of the game is $Y_n = \text{Min}(X_1, \ldots, X_n)$, where $X_1, \ldots, X_n$ are i.i.d. exponential with mean $(n-1)(V_{n-1} - V_n)$. Hence Y_n is exponential with mean $\frac{n-1}{n}(V_{n-1} - V_n)$ and, since each player has probability $1/n$ of collecting V_n, the mean reward to a player at the beginning of a game is

$$\frac{1}{n} V_n + \frac{n-1}{n} V_{n-1} - \frac{n-1}{n}(V_{n-1} - V_n) = V_n \ .$$

We now prove that the stated strategy is an ESS. The players select times $X_1, X_2, \ldots, X_n$ in some fashion, and define $W_1 = \text{Min}(X_2, \ldots, X_n)$. The reward to player 1 is thus $V_n - X_1$ if $X_1 < W_1$ and (using the inductive assumption) $V_{n-1} - W_1$ if $X_1 > W_1$. But this is equivalent to a 2-person WA with the single reward $V_{n-1} - V_n$; hence, in order that the mean payoff be constant for all x in the support of X_1, W_1 has an exponential distribution with mean $V_{n-1} - V_n$. Thus

$$\exp\left(\frac{-t}{V_{n-1} - V_n}\right) = P(W_1 > t) = P(X > t)^{n-1} \ ,$$

showing that the only ESS candidate for X is the exponential distribution with mean

$(n-1)(V_{n-1} - V_n)$; call this G.

As in the investigation of model A, we must prove that the expression I in (5) is positive unless $H = G$. Using $T(x)$ in the same manner as that analysis, write $W = \mathrm{Min}\,(Y, X_3, \ldots, X_n)$ so that here

$$T(x) = (V_n - x)(1 - H(x))(1 - G(x))^{n-2} - \int_0^x (V_{n-1} - t)\, d((1 - H(t))(1 - G(t))^{n-2})$$

i.e.

$$T(x) = (V_n - V_{n-1})(1-H)(1-G)^{n-2} + V_{n-1} - \int_0^x (1-H)(1-G)^{n-2}\, dt . \tag{11}$$

Since $I = \int_0^\infty T(x)(g(x) - h(x))\, dx$, from (11) we find

$$I = \int_0^\infty \{(V_n - V_{n-1})(1-H)(1-G)^{n-2}(g-h) + (G-H)(1-H)(1-G)^{n-2}\}\, dx$$

$$= \int_0^\infty f(x, H, h)\, dx, \ \text{say, where } G = G(x) = 1 - \exp\left(\frac{-x}{(n-1)(V_{n-1} - V_n)} \right) .$$

Once again, Euler's Equation is shown to have the unique solution $H = G$. Using the same notation as in the analysis of Model A, $P \equiv 0$ and

$$Q = 1/2(2(1-G)^{n-2} + (n-2)(V_n - V_{n-1})g(1-G)^{n-3})$$

$$= 1/2(1-G)^{n-2}, \ \frac{n}{n-1} \geq 0 .$$

Thus $H = G$ is a minimum of I, which completes the proof.

Suppose some or all of the inequalities $V_{i+1} < V_i$ are replaced by $V_{i+1} \leq V_i$. If $V_n = V_{n-1}$, the formal interpretation of the distribution of X in Theorem 4 is that X corresponds to the strategy "Quit immediately". If we make the same "reasonable" interpretation of what would happen in practice if several players each decide to quit immediately as in Model B, i.e. that one of these n, chosen at random, quits with reward V_n leaving the others to contest a contest with rewards $(V_{n-1}, \ldots, V_1)$, it is clear that no other strategy can do better. Thus, with this formal interpretation of X in Theorem 4, the conclusions of Theorem 4 also hold when $V_n \leq V_{n-1} \leq \cdots \leq V_1$.

We now see that Model B, with rewards $(0,0,\ldots,0,V)$ to the n participants, also fits this pattern.

Matters are rather different when the rewards $\{V_i\}$ do not become available in the order of increasing value. Consider the two-person WA with rewards (V_2,V_1), but $V_2 > V_1$. It is clear that the optimal strategy is to quit immediately; call this strategy Q. Since both players use Q, the mean payoff is $\bar{V} = (V_1 + V_2)/2$ and the game leads to the same policies as if the two payoffs were $(\bar{V},\bar{V})$.

For a three-person WA with $V_3 \geq V_2$, then to play Q at the first stage is optimal, whatever the order of V_2 and V_1. But if $V_3 < V_2$ and $V_2 > V_1$, the optimal policy at the first stage will depend on whether $V_3 < \bar{V}$ $(= (V_1 + V_2)/2)$ or $V_3 \geq \bar{V}$. If $V_3 \geq \bar{V}$, clearly Q is best but, if $V_3 < \bar{V}$, the optimal policy is the same as if the rewards were $(V_3,\bar{V},\bar{V})$, i.e. an ESS is to use an exponential distribution with mean $2(\bar{V} - V_3)$.

This combination of the use of Q and certain exponential distributions turn out to be characteristic of a contest in which the rewards are not presented in order of increasing value.

Theorem 5. Given any set of rewards for a Model C WA, presented in the order $(V_n,V_{n-1},\ldots,V_1)$, we can construct a set of rewards in the order $W_n \leq W_{n-1} \leq \cdots \leq W_1$ for a Model C WA, whose unique ESS is described in Theorem 4 (and the paragraph following its proof). Optimum play for both games is the same.

Proof. In the preamble to this theorem, we showed how to construct the $\{W_i\}$ if $n = 2$ or 3. Suppose, given $(V_{n-1},\ldots,V_1)$ we have successfully constructed suitable $(W_{n-1}^{(n-1)},\ldots,W_1^{(n-1)})$. If $V_n \leq W_{n-1}^{(n-1)}$, define $W_n^{(n)} = V_n$ and $W_i^{(n)} = W_i^{(n-1)}, 1 \leq i \leq n-1$; (case 1). But if $V_n > W_{n-1}^{(n-1)}$, define

$$W_n^{(n)} = \frac{1}{n} V_n + \frac{n-1}{n} W_i^{(n-1)} \quad \text{and} \quad W_{n-r}^{(n)} = W_{n-r}^{(n-1)} + \frac{1}{n}(V_n - W_{n-1}^{(n-1)}) \quad \text{for}$$

$r = 1,2,\ldots,n-1$; (case 2). In either case it is clear that $W_n^{(n)} \leq \cdots \leq W_1^{(n)}$. In case 1, it is easy to see that if optimum play for $(V_{n-1},\ldots,V_1)$ is the same as optimum play for $(W_{n-1}^{(n-1)},\ldots,W_1^{(n-1)})$, the same is true for the two games with n rewards. In case 2, since $W_n^{(n)} = W_{n-1}^{(n)}$, the policy for $(W_n^{(n)},\ldots,W_1^{(n)})$ is to quit immediately at the first stage, leaving $n-1$ players to contest the game with rewards $(W_{n-1}^{(n)},\ldots,W_1^{(n)})$. But $W_i^{(n)} - W_{i-1}^{(n)} = W_i^{(n-1)} - W_{i-1}^{(n-1)}$ $(2 \leq i \leq n-1)$, so the remaining play is identical with the play for $(W_{n-1}^{(n-1)},\ldots,W_1^{(n-1)})$, with value $W_{n-1}^{(n-1)}$ to each player. But in case 2, optimum play in $(V_n,W_{n-1}^{(n-1)},\ldots,W_1^{(n-1)})$ is to quit immediately and move into the game $(W_{n-1}^{(n-1)},\ldots,W_1^{(n-1)})$; so the value of the whole game is

$$\frac{1}{n} V_n + \frac{n-1}{n} W_{n-1}^{(n-1)} = W_n^{(n)} = W_{n-1}^{(n)}.$$

The construction of $W_n \leq W_{n-1} \leq \cdots \leq W_1$, from an arbitrarily ordered $(V_n, V_{n-1}, \ldots, V_1)$ is similar to the method of backwards induction used in dynamic programming: at any stage n, we assume that the optimum policy has been found for all stages up to $n-1$, and then use this policy to construct one further optimal step. An illustrative example follows.

Example 1. Consider a 7-player Model C WA with rewards in the order (19,5,12,7,9,6,12). Theorem 5 successively constructs W-values as (6,12); (7,7,13); (7,7,7,13); (8,8,8,8,14); (5,8,8,8,8,14) and (7,7,10,10,10,10,16). The optimum play at each stage of the game is $Q, E(15), Q, Q, Q, E(6)$, (where $E(\mu)$ means an exponential random variable with mean μ), and the value of the game to each participant is 7.

With the same rewards, but in increasing order (5,6,7,9,12,12,19), the value of the game is reduced to 5, and optimum play is, successively $E(6)$, $E(5)$, $E(8)$, $E(9)$, Q, $E(7)$. It is clear that, from the viewpoint of the population, who would wish the value of the game to be maximised, this is achieved (for a given set of reward values) when they are available in decreasing order of value; and minimised when in increasing order.

Model D. This differs from Model C only in that, as in Model A, players may not change their strategy as opponents drop out. Rewards are available in the order $(V_n, V_{n-1}, \ldots, V_1)$ and players decide, at the outset, how long to wait, they then collect the reward available at the time they quit.

Any possible ESS will be some probability distribution G that governs the choice of the time to wait. If z is a point of increase of G, then $E(z)$, the mean reward to a user of z when $(n-1)$ opponents use G, must be constant. Now

$$E(z) = \sum_{r=0}^{n-2} (V_{n-r} - z)\binom{n-1}{r} G(z)^r (1 - G(z))^{n-r} + \int_0^z (V_1 - y) d(G(y)^{n-1}) . \quad (12)$$

The equation $E'(z) = 0$ reduces to

$$1 - G^{n-1} = (n-1)g \sum_{r=0}^{n-2} (V_{n-r-1} - V_{n-r})\binom{n-2}{r} G^r (1-G)^{n-2-r} \quad (13)$$

where, as before, $G \equiv G(z)$, and $g \equiv g(z) = G'(z)$.

It is easy to use (13) to deduce that, for some choices of $\{V_i\}$, the formal solution G of (13) is *not* a distribution function. Write $V_{n-r-1} - V_{n-r} = C_{n-r-1}$, and consider, as an example, the case $n = 3$. Here (13) is just

$$\frac{1 - G^2}{2g} = c_2(1-G) + c_1 G ; \quad (14)$$

the conditions $0 \leq G \leq 1$ and $g \geq 0$ are clearly violated if either c_1 or c_2 is negative, which arises, for example, if $V_3 < V_2 > V_1$. The formal solution of (14) arises via

$$\int dx = \int \frac{2(c_2 + (c_1 - c_2)G)}{1 - G^2} \, dG$$

$$= \int \left(\frac{c_1}{1-G} + \frac{2c_2 - c_1}{1+G}\right) dG$$

i.e.

$$x = (2c_2 - c_1)\ln(1+G) - c_1\ln(1-G) . \tag{14A}$$

Provided both $c_1 > 0$ and $c_2 > 0$, then (14), (14A) show that G is a distribution function, whose corresponding density has support $[0,\infty)$.

Returning to the general case (13), suppose G is a genuine distribution function. We now have to consider whether it represents an ESS, by the method and notation used for Models A and C. When pure strategy x is played against $(n-2)$ opponents who use G and one who uses H, the mean reward is $T(x)$,

$$T(x) = (1-H) \sum_{r=0}^{n-2} (V_{n-r} - x)\binom{n-2}{r} G^r (1-G)^{n-2-r} \tag{15}$$

$$+ H \sum_{r=1}^{n-1} (V_{n-r} - x)\binom{n-2}{r-1} G^{r-1}(1-G)^{n-1-r} + \int_0^x H(y) G^{n-2}(y) \, dy .$$

As before, use Euler's Equation on $\int_0^\infty T(x)(g(x) - h(x)) \, dx$ to get

$$0 = 1 - 2HG^{n-2} + G^{n-1} - (n-2)g\{(1-H) \sum_{r=0}^{n-3} c_{n-1-r}\binom{n-3}{r} G^r (1-G)^{n-3-r} \tag{16}$$

$$+ H \sum_{r=0}^{n-3} c_{n-2-r}\binom{n-3}{r} G^r (1-G)^{n-3-r}\} - g \sum_{r=0}^{n-2} c_{n-1-r}\binom{n-2}{r} G^r (1-G)^{n-2-r} .$$

Collect together the terms involving H in (16), and use (13) to eliminate g. This leads to

$$H \left\{2(n-1)G^{n-2} + \frac{(n-2)(1 - G^{n-1})}{U} \sum_{r=0}^{n-3} (c_{n-2-r} - c_{n-1-r})\binom{n-3}{r} G^r (1-G)^{n-3-r}\right\}$$

$$= (n-1)(1- G^{n-1}) - \frac{(n-2)(1 - G^{n-1})}{U} \sum_{r=0}^{n-3} c_{n-1-r} \binom{n-3}{r} G^r (1-G)^{n-3-r}$$

$$- \frac{(1 - G^{n-1})}{U} \sum_{r=0}^{n-2} c_{n-1-r} \binom{n-2}{r} G^r (1-G)^{n-2-r}$$

where

$$U = \sum_{r=0}^{n-2} c_{n-1-r} \begin{bmatrix} n-2 \\ r \end{bmatrix} G^r (1-G)^{n-2-r} . \tag{17}$$

After multiplying through by U and simplifying, we find

$$0 = (H-G) \sum_{r=0}^{n-2} c_{n-1-r} \begin{bmatrix} n-2 \\ r \end{bmatrix} G^{r-1}(1-G)^{n-3-r} \tag{18}$$

$$\{r - (n-2)G + (2n-r-2)G^{n-1} - nG^n\}$$

Hence, the only solution of Euler's Equation for H is $H = G$, where G satisfies (13) which, using (17), can be written

$$1 - G^{n-1} = (n-1)gU . \tag{13A}$$

The second order conditions for a minimum, on the expression J, lead to $P = 0$ and

$$2Q = 2G^{n-2} - \frac{d}{dx} \{ - \sum_{r=0}^{n-2} c_{n-r-1} \begin{bmatrix} n-2 \\ r \end{bmatrix} G^r (1-G)^{n-2-r} \}$$

which, when evaluated, leads to

$$2(n-1)QU = \sum_{r=0}^{n-2} c_{n-1-r} \begin{bmatrix} n-2 \\ r \end{bmatrix} G^{r-1}(1-G)^{n-3-r}\{r - (n-2)G - nG^n + (2n-r-2)G^{n-1}\} .$$

For models A and C, it was clear that the corresponding Q was non-negative, and thus G was seen to represent an ESS. But (17) and (18) show that, for general $\{c_k\}$, to see whether $J(\alpha)$ is a non-negative function is a complex matter.

Example 2. Take $n = 3$ with $c_1 > 0$, $c_2 > 0$. We have seen that G is a distribution function, and (18) reduces to

$$4Q\{c_2(1-G) + c_1 G\} = c_2(1-G)(3G-1) + c_1(1+3G^2) .$$

If $c_1 \geq c_2$, clearly $Q \geq 0$, and so G is an ESS. In particular, if $c_1 = c_2 = c$, equation (14) shows that the unique ESS is

$$G(x) = \frac{\exp(x/c) - 1}{\exp(x/c) + 1} \ (x \geq 0) .$$

However, suppose $c_2 = 3$ and $c_1 = 2$. Then Q is obtainable from

$$4Q(3-G) = -1 + 12G - 3G^2$$

so plainly the sufficient condition $J(\alpha) \geq 0$ is not satisfied. We can find G from (14A), which is $z = 4\ln(1 + G) - 2\ln(1 - G)$, i.e.

$$G(2) = (\frac{e^z}{4} + 2e^{z/2}) - 1 - 1/2\, e^{z/2}\, (z \geq 0)\ .$$

This function G is a distribution function, and it is the sole candidate for an ESS. But it is not an ESS, and this contest has no ESS.

To see this, let H be the pure strategy concentrated at some point α (to be chosen later), and take the rewards to have values 0 , 3 and 5. Using the notation of expression (4A), we know from the way G was constructed, that $E(H;G^{(2)}) = E(G;G^{(2)})$. Thus (see (5)), we seek to choose α so that

$$E(H;H,G) > E(G;H,G)\ ;$$

when this occurs, strategy H can invade a population of G-users, hence G is not an ESS.

By considering whether or not the G-user waits longer than α, it is easy to see that

$$E(H;H,G) = (1 - G(\alpha))[1/2\ .\ 0 + 1/2\ .\ 3 - \alpha] + G(\alpha)[1/2\ .\ 3 + 1/2\ .\ 5 - \alpha]$$

i.e.

$$E(H;H,g) = 2.5\,G(\alpha) + 1.5 - \alpha\ . \tag{19}$$

Similarly, by conditioning on the choice of strategy of the G user faced with H,G, we get

$$E(G;H,G) = \int\limits_0^\alpha g(z)[1-G(z))(0-z) + G(z)(3-z)]\,dz$$

$$+ \int\limits_\alpha^\infty g(z)\{(1 - G(z))(3-z) + G(\alpha)(5-\alpha) + \int\limits_\alpha^z g(y)(5-y)\,dy\}\,dz\ .$$

After some manipulations, this can be reduced to

$$E(G;H,G) = 4 - 3G(\alpha) + 0.5G(\alpha)^2 - \int\limits_0^\alpha (1-G(z))\,dz - \int\limits_\alpha^\infty (1-G(z))^2\,dz\ . \tag{20}$$

From (19) and (20)

$$E(H;H,G) - E(G;H,G) = 5.5\,G(\alpha) - 0.5G(\alpha)^2 - 2.5 - \alpha + \int\limits_0^\alpha (1-G(z))\,dz \tag{21}$$

$$+ \int\limits_\alpha^\infty (1-G(z))^2\,dz\ .$$

But, for a range of values of α, the right side of (21) is positive. For example, when $G(\alpha) = 3/4$, which corresponds to $\alpha = 5$, the right side of (21) (even neglecting the last term) is about 1.3.

Example 3. Suppose $n = 4$ so that (13) is

$$1 - G^3 = 3g(c_3(1-G)^2 + 2c_2G(1-G) + c_1G^2) \tag{22}$$

and (18) simplifies to

$$3Q[c_3(1-G)^2 + 2c_2G(1-G) + c_1G^2] = -c_3(1-G)^3(1+2G) \tag{23}$$

$$+ c_2(1-G)(1-G-G^2+4G^3) + c_1G(1+2G^3) \ .$$

Sufficient conditions for the solution G of (22) to be a distribution function are that each of c_1, c_2, c_3 is positive. And, in this case, the right side of (23) is positive for all G if $c_2 \geq c_3$ and $c_1 \geq c_3$. Thus $Q \geq 0$, so G is an ESS.

Summarising. Sufficient conditions for the solution G of (22) to be the unique ESS of this game are that $c_1 \geq c_3$, $c_2 \geq c_3$, $c_3 > 0$.

Example 4. Suppose $c_{n-1} = c_{n-2} = \cdots = c_1 = c$ in (13). Then (13) is just

$$1 - G^{n-1} = c(n-1)g$$

which is solved via

$$\int \frac{dG}{1-G^{n-1}} = \frac{x}{c(n-1)} \ . \tag{24}$$

Equation (18) also simplifies nicely, reducing eventually to

$$2Q(n-1) = 2(n-1)G^{n-2}$$

which shows that the solution of (24), which is a distribution function is also the unique ESS of this model (provided, of course, that $c > 0$).

This might be called the *linear* model, as the rewards $\{V_n\}$ form an increasing arithmetic progression.

We summarize our results for Model D as a Theorem.

Theorem 6. In Model D, the only candidate for an ESS is the function G satisfying

$$1 - G^{n-1} = (n-1)g \sum_{r=0}^{n-2} c_{n-r-1} \binom{n-2}{r} G^r(1-G)^{n-2-r} \tag{13}$$

where $c_k = V_k - V_{k+1}$.

(a) For some values of $\{c_k\}$, the solution of (13) is not even a distribution function, and no ESS exists.

(b) For other values $\{c_k\}$, the solution of (13) is a distribution function, but it is not an ESS, and there is no ESS.

(c) If $0 < c_{n-1} = c_{n-2} = \cdots = c_1$, the solution of (13) is an ESS.

(d) If $0 = c_{n-1} = \cdots = c_2$, $c_1 = V$ the solution of (13) is that G appropriate for Model A.

3. References

Abramowitz, M. and Segun, I.A. (1965). Handbook of Mathematical Functions. Dover, N.Y.

Bishop, D.T. and Cannings, C. (1976). Models of animal conflict. Adv. Appl. Prob., **8**, 616-621.

Bishop, D.T. and Cannings, C. (1978). A generalised war of attrition. J. Theoret. Biol. **70**, 85-125.

Bishop, D.T. and Cannings, C. (1986). Ordinal conflicts with random rewards. J. Theoret. Biol. **122**, 225-230.

Gelfand, I.M. and Fomin, S.V. (1963). Calculus of Variations. Prentice Hall.

Haigh, J. and Rose, M.R. (1980). Evolutionary game auctions. J. Theoret. Biol., **85**, 381-97.

Maynard Smith, J. (1974). The theory of games and the evolution of animal conflicts. J. Theoret. Biol., **47**, 209-21.

Maynard Smith, J. and Parker, G. (1976). The logic of asymmetric contents. Anim. Behav. **24**, 159-75.

Palm, G. (1984). Evolutionary Stable Strategies and game dynamics for n-person games. J. Math. Biol., **19**, 329-334.

Acta Applicandae Mathematicae **14** (1989), 75–89.

Mutation-Selection Models in Population Genetics and Evolutionary Game Theory

Reinhard Bürger

Institut für Mathematik
Universität Wien
Strudlhofgasse 4
A-1090 Wien, Austria

AMS Subject Classification (1980): 92A10
Key words: mutation-selection equation, continuum of alleles, gradient systems

1. Introduction

The investigation of models including the effects of selection and mutation has a long history in population genetics, which can be traced back to Fisher, Haldane and Wright. The reason is that selection and mutation are important forces in evolution. Selection favours an optimal type and tends to eliminate genetic variation, in particular in quantitative traits. Mutation, on the other hand, is the ultimate source of genetic variability. Traditionally, models with only two alleles per locus have been treated. At the end of the fifties the first general results for multi-allele models with selection but without mutation were proved. In particular, conditions for the existence of a unique and stable interior equilibrium were derived and Fisher's *Fundamental Theorem of Natural Selection* was proved to be valid (e.g. Mulholland and Smith, 1959; Scheuer and Mandel, 1959; Kingman, 1961a). It tells that in a one-locus multi-allele diploid model mean fitness always increases. It is well known now that in models with two loci or more this is wrong in general.

The first general results for models with selection and mutation and many alleles segregating at a gene locus have been obtained in the seventies. Moran (1976) demonstrated existence, uniqueness and global stability of an equilibrium for haploid mutation-selection equations in discrete time. Even earlier, Thompson and McBride (1974) had derived the solutions of a system of differential equations occuring in Eigen's theory of the evolution of macromolecules that is formally equivalent to the haploid mutation-selection model.

It is well known that mutation may – in principle - lead to a huge number of possible alleles at a locus. To cope with this, models have been introduced allowing for an infinite number of possible alleles. One of the most influential among these is the *continuum-of-alleles model* proposed by Crow and Kimura (1964). In this model it is assumed that an allele with average effect x on a quantitative trait may mutate to an allele with effect $x+y$ according to a certain probability distribution. Another type of model in which alleles are labelled by integers $0, \pm1, \pm2, \pm3, \ldots$ was suggested by Bulmer (1971) and by Ohta and Kimura (1973) to investigate the number of electrophoretically detectable alleles. In this model an allele A_i may mutate to A_{i-1} or A_{i+1} with equal probability. It is called the *stepwise mutation model*. Such models became increasingly important during recent years and have been studied extensively (see below). An intrinsic assumption to these models is that the population reproduces asexually.

True diploid models are much more complicated. For example Akin (1979) and Hofbauer (1985) have shown that stable limit cycles may exist. On the other hand, Hadeler (1981) and, more generally, Hofbauer (1985) proved that a stable polymorphism in the pure selection model remains stable, if mutation is added in such a way that mutation rates depend only on the resulting allele.

In section 2 of the present paper I will review the state of the art of general haploid mutation-selection models and discuss some applications. Section 3 contains a review of results for diploid populations. In section 4 replicator equations describing the dynamics of mixed strategies will be considered. A general theorem is proved assuring that an ESS remains stable if mutation is introduced into the dynamic equations. Its proof is based on the investigation of spectral properties of positive irreducible matrices and uses a theorem of Kingman (1961b) on the convexity of the spectral radius.

2. Selection and mutation in haploid populations

This section presents a general model for mutation and selection in haploid populations, its analysis and the relation to earlier work. I begin with settling the notation and the general assumptions.

The model assumes an effectively infinite population that reproduces asexually and has overlapping generations. In other terms a deterministic haploid model in continuous time is considered. Individuals are characterized by their type x, where x is a vector in a locally compact subset $M \subseteq \mathbb{R}^k$. M is the state space and is endowed with a positive σ-finite measure ν. Important special cases are $M = \{1, 2, \ldots, n\}$, $M = \{0, \pm1, \pm2, \ldots\}$ and $M = \mathbb{R}$. In the first and second case each $x \in M$ can be interpreted as a (possible) allele at a given gene locus. In the third case each $x \in M$ may be considered as an allelic effect on a quantitative trait. $p(x, t)$ denotes the normalized density of type x in the population at time t. $p(t): x \mapsto p(x, t)$ denotes the corresponding element in $L^1(M, \nu)$.

This is the Banach space of absolutely ν-integrable complex-valued functions on M, i.e., $f \in L^1(M)$ if $\|f\| = \int_M |f(x)| \, d\nu(x) < \infty$. Hence $p(t)$ is positive and $\|p(t)\| = 1$.

Below a continuous time model is used and therefore fitness is assumed to be measured in Malthusian parameters. Fitness of type x is denoted by $m(x)$. It is assumed that $m: M \to \mathbb{R}$ is measurable and that $m(x) \leq \text{const.}(\nu\text{-a.e.})$. The mutation term will be denoted by $u(x,y) \geq 0$, i.e. $u(x,y)dt$ is the fraction of individuals of type x originating through mutation from individuals of type y during the time intervall dt. It is assumed that $u: M \times M \to \mathbb{R}_+$ is measurable and that $\mu(x) = \int_{M \setminus \{x\}} u(y,x) \, d\nu(y) < 1$, for (almost) all x. The latter condition just tells that the mutation rates are less than one. Usually $\mu(x)$ is very small.

Employing standard modelling techniques from population genetics (cf. Kimura, 1965) the differential equation describing the dynamic behaviour of type densities $p(x,t)$ in a haploid population is derived to

$$\frac{\partial p(x,t)}{\partial t} = [m(x) - \bar{m}(t)]p(x,t) + \int_M u(x,y)p(y,t) \, d\nu(y) - \mu(x)p(x,t), \qquad (1)$$

where $\bar{m}(t) = \int_m m(x)p(x,t) \, d\nu(x)$ is the mean fitness of the population at time t. In order to prove a general result concerning the dynamic and equilibrium properties of equation (1), methods from functional analysis are needed. To this aim the following operators on $L^1(M)$ with their domains are defined

$$\begin{aligned}
Tf(x) &= w(x)f(x), & D(T) &= \{f \in L^1(M): wf \in L^1(M)\}, \\
Uf(x) &= \int_M u(x,y)f(y) \, d\nu(y), & D(U) &= L^1(M), \\
A &= T - U, & D(A) &= D(T).
\end{aligned}$$

Here $w = -m + \mu - \text{ess inf}(-m + \mu)$. This assures that $w: M \to \mathbb{R}_+$ is measurable, ess inf $w = 0$ and that $(w+1)^{-1}$ is essentially bounded.

U is a bounded positive operator on $L^1(M)$, but T and A are not. However, T and A are closed operators and $-T$ and $-A$ generate positive analytic semigroups. Using these operators the partial differential equation (1) can be written as an ordinary differential equation on the Banach space $L^1(M)$:

$$\dot{p}(t) + Ap(t) = p(t) \int_M (Ap)(y,t) \, d\nu(y). \qquad (2)$$

The stationary or equilibrium solutions of (2) are precisely those functions in $L^1(M)$ that satisfy

$$Af_\alpha = -\alpha f_\alpha, \qquad f_\alpha \in D(A) \qquad (3)$$

for some complex number α, since $Af_\alpha = -\alpha f_\alpha$ implies $-\alpha = \int Af_\alpha \, d\nu(x)/\int f_\alpha \, d\nu(x)$. In the present context we are only interested in the existence of positive solutions of (2) normalized such that $\int f_\alpha \, d\nu(x) = 1$.

Eq. (3) is an eigenvalue problem for the unbounded operator A. For the present problem only positive eigenfunctions are of interest. To solve it the following family of operators on the Banach space $F = \{f \in D(T)\}$ with norm $\|f\|_F = \|f\| + \|Tf\|$ is introduced

$$K_\alpha f(x) = U(T + \alpha)^{-1} f(x) = \int_M \frac{u(x, y)}{w(y) + \alpha} f(y) \, d\nu(y), \quad \alpha > 0. \tag{4}$$

Each K_α is a bounded positive operator on F. The trick is that if $\alpha > 0$ then f_α is an eigenfunction of A corresponding to an eigenvalue $-\alpha$, i.e. (3) holds, if and only if $g_\alpha = (T + \alpha) f_\alpha$ is an eigenfunction of K_α corresponding to the eigenvalue 1, i.e.

$$K_\alpha g_\alpha = g_\alpha. \tag{5}$$

To show that a unique α_0 exists satisfying (5) (this implies that (3) has a unique positive solution f_{α_0}) it is shown that the function $\alpha \mapsto r(K_\alpha)$, where $r(K_\alpha)$ is the spectral radius of K_α, is strictly monotone decreasing, continuous and satisfies $\lim_{\alpha \to \infty} r(K_\alpha) = 0$ and $r(K_\alpha) > 1$ if α is small.

To this aim three additional conditions are needed, namely (i), (ii) and (iii) in Theorem 1 below.

Theorem 1. *Assume that the following conditions are satisfied.*

(i) There exists some $n \geq 1$ such that K_α^n is compact for all $\alpha > 0$.

(ii) U is irreducible, that is, if $S \subseteq M$, $\nu(S) > 0$ and $\nu(M \setminus S) > 0$ then

$$\int_{M \setminus S} \int_S u(x, y) \, d\nu(x) \, d\nu(y) > 0 \tag{6}$$

(iii) There exists $x_0 \in M$ with $w(x_0) = 0$, a neighbourhood I of x_0 and a constant $u_0 > 0$ such that

$$u(x, y) \geq u_0, \qquad (x, y) \in I \times I \tag{7}$$

and

$$u_0 \lim_{\alpha \downarrow 0} \int_I (w(y) + \alpha)^{-1} \, d\nu(y) > 1. \tag{8}$$

Then a uniquely determined positive equilibrium solution $\hat{p}$ of (1) (and (2)) exists. It is the normalized eigenfunction $f_{\alpha_0}/\|f_{\alpha_0}\|$ corresponding to the unique eigenvalue $-\alpha_0$ of (3) that satisfies (5). $\hat{p}$ is even strictly positive ν - a.e. Moreover, $\hat{p}$ is globally asymptotically *stable in the sense that for any $p_0 \geq 0$ with $\int_M p_0(x)\,d\nu(x) = 1$ the uniquely determined solution $p(x,t)$ of (1) (or of (2)) with $p(x,0) = p_0(x)$ satisfies*

$$\lim_{t \to \infty} \int_M |p(x,t) - \hat{p}(x)|\,d\nu(x) = 0. \tag{9}$$

Remark. Condition (i) ensures that the spectral radius $r(K_\alpha)$ is an eigenvalue of K_α of finite multiplicity. Condition (i) is met, for example, if

$$\int \operatorname{ess\,sup}_{y \in M} \frac{u(x,y)}{w(y) + 1}\,d\nu(x) < \infty \tag{10}$$

or if M is a subgroup of $\mathbb{R}^k$ and the following holds

$$u(x,y) = \tilde{u}(x - y), \quad \tilde{u} \in L^1(M) \quad \text{and} \quad (w + 1)^{-1} \text{vanishes at infinity}. \tag{11}$$

(10) implies compactness of K_α^2, and (11) compactness of K_α. Condition (ii) ensures uniqueness and strict positivity of $\hat{p}$. Condition (iii) is a kind of cusp condition telling that at x_0 the function w must not have a cusp. It implies that $r(K_\alpha) > 1$ as $\alpha \to 0$ If (8) is not satisfied, I argue that an atom of probability will occur at $\{x_0\}$. For related models this was shown by Kingman (1977, 1978). The complete proof of Theorem 1 may be found in Bürger (1988a, see Theorem 3.5), where a general theorem on perturbations of generators of positive semigroups was proved and applied to mutation-selection models. A result analogous to Theorem 1 for a discrete time model was proved in Bürger (1988b). In that case a stronger assumption than irreducibility, namely an assumption related to primitivity, is needed.

Next, we consider some applications and special cases. Suppose that M is a discrete, countable set with measure ν normalized such that $\nu(x) = 1$, $x \in M$. Writing $p_i(t)$ instead of $p(x,t)$ with $i = x$, $m_i = m(x)$, $u_{ij} = u(i,j)$ if $i \neq j$ and putting $u_{ii} = 1 - \sum_{j \neq i} u_{ji}$ equation (1) becomes

$$\dot{p}_i = p_i(m_i - \bar{m}) + \sum_{j=1}^n u_{ij} p_j - p_i. \tag{12}$$

Now u_{ij} denotes just the mutation rate from allele A_j to allele A_i.

If M is finite this is the classical haploid one-locus multi-allele model with mutation and selection (Crow and Kimura, 1970). Thompson and McBride (1974) have solved eq. (12) exactly for finite M, by transforming it to a system of linear differential equations. However, their equations had a different biological background. They describe the evolution of self-reproducing macromolecules when replication errors occur (see Eigen,

1971). Performing a perturbation treatment for small mutation rates they established an approximation for the unique positive equilibrium solution. Later, but independently, Moran (1976) considered a corresponding discrete time model and proved existence, uniqueness and global stability of a strictly positive equilibrium solution under the assumption that the mutation matrix (u_{ij}) is primitive. He used Perron-Frobenius theory of positive matrices. Of course Theorem 1 implies the same result for eq. (12), since for finite M conditions (i) and (iii) are automatically satisfied and irreducibility is sufficient for continuous time.

If $M = \{0, \pm 1, \pm 2, \ldots\}$ with mutation occuring from allele A_i to A_{i-1} and A_{i+1} at the rate $\mu/2$ each, the so-called *stepwise mutation model* is obtained. It was suggested by Bulmer (1971) and, more explicitly, by Ohta and Kimura (1973, 1975) to estimate the number of electrophoretically detectable alleles and the variability maintained in a population. Moran (1976, 1977) and Kingman (1977) proved the existence of a stationary distribution under more general conditions on the fitness function than I can, but convergence to this stationary distribution occurs only from a restricted set of initial distributions. These authors used a discrete time model. From Theorem 1 existence, uniqueness and global stability of an equilibrium follow immediately for the continuous time model (12), if conditions (i) and (ii) are met. (8) is always satisfied, if $u(x_0, x_0) > 0$ where $w(x_0) = 0$. This is clearly fulfilled in all biological applications.

If $M = \mathbb{R}$, then (1) describes the evolution of densities of types in the so-called *continuum-of-alleles* model of Crow and Kimura (1964). Continuum-of-alleles models play an important role in quantitative genetics, in particular in connection with the problem how much heritable variation exhibited by metric traits can be maintained through a balance between stabilizing selection and mutation. Kimura (1965) was interested in stabilizing selection and assumed $m(x) = -sx^2$, $s > 0$ the selection coeficient. As mutation function he chose

$$u(x, y) = \mu(2\pi\alpha^2)^{-1/2} \exp\left\{\frac{-(x-y)^2}{2\alpha^2}\right\}, \tag{13}$$

where μ denotes the mutation rate (assumed to be equal for all types). Hence, in the present notation $\mu(x) = \mu$. He performed a diffusion approximation and asserted that the equilibrium distribution be normal with mean zero and variance

$$\hat{\sigma}^2(G) = \alpha\sqrt{\mu/2s}, \tag{14}$$

the G standing for Gaussian approximation. This assertion, however, is only true for his approximating diffusion equation. More sophisticated approximation techniques were applied by Fleming (1979), Nagylaki (1984) and Turelli (1984) to figure out the variance of the equilibrium density, which is the parameter of biological interest. Turelli chose

$$u(x, y) = \mu(2\pi\alpha^2)^{-1/2} \exp\left\{\frac{-x^2}{2\alpha^2}\right\}, \tag{15}$$

for all y. This simplifies (1) considerably, since the integral on the right hand side of (1) is then simply replaced by the term on the right hand side of (15). The resulting equation is called the house-of-cards model, which has been introduced in a slightly different fashion by Kingman (1977, 1978). Turelli (1984) showed that the choice (15) for u may be used to derive an excellent approximation for the variance of the equilibrium distribution of (1) with u given by (13), if the variance of mutational effects α^2 is much larger than the equilibrium variance. His house-of-cards approximation for the equilibrium variance is

$$\hat{\sigma}^2(HC) = \mu/s, \tag{16}$$

which is independent of α^2. For a discussion of the relation between the Gaussian and the house-of-cards model the reader is referred to Turelli (1986). Contrary to Kimura, however, Fleming, Nagylaki and Turelli considered a discrete time model. But their analyses carry over to continuous time.

None of these authors has shown that a unique equilibrium solution of (1) in fact exists, nor has investigated the stability properties. A first proof for the special case considered by Kimura was given by Bürger (1986) using the theory of selfadjoint operators on $L^2(\mathbb{R})$. Theorem 1 provides a general result on equilibrium and stability properties of equation (1). It is easily seen that the fitness function $m(x) = -sx^2$ and the mutation functions given in (13) and (15) satisfy all the conditions of Theorem 1. For the discrete time version of (1) existence and global stability of a uniquely determined strictly positive equilibrium solution was proved by Bürger (1988b). Finally, from Theorem 1 an upper bound for the equilibrium variance of the stationary distribution can be derived if fitness is given by $m(x) = -sx^2$ and the mutation distribution satisfies $\int u(y, x)\, d\nu(y) = \mu$ for all x. This upper bound is just μ/s, which agrees with the house-of-cards approximation. These models have important applications in quantitative genetics, in particular in connection with the problem how much genetic variation can be maintained by a balance between mutation and stabilizing selection (see Turelli, 1984 and Bürger, Wagner, Stettinger, 1988, for example).

3. Selection and mutation in diploid populations

This section summarizes some of the most important results on mutation-selection models in diploid populations. Throughout, only the case where a finite number of alleles $A_1, \ldots, A_n$ at a gene locus exist is treated. The dynamic properties are much more complicated than in haploid models. Following Crow and Kimura (1970) the differential equations describing the evolution of allele frequencies p_i under selection and mutation in a random mating population are given by

$$\dot{p}_i = p_i(m_i - \bar{m}) + \sum_{j=1}^{n}(u_{ij}p_j - u_{ji}p_i). \tag{17}$$

Here u_{ij} is the mutation rate from A_j to A_i (for $i \neq j$), satisfying $u_{ij} \geq 0$ and $\sum_{i=1}^{n} u_{ij} = 1$ for all $j = 1, \ldots, n$. Furthermore, $m_i = \sum_{j=1}^{n} m_{ij}p_j$ is the marginal fitness of allele A_i and $m_{ij} = m_{ji}$ is the fitness of the genotype A_iA_j. Finally, $\bar{m} = \sum_{i,j=1}^{n} m_{ij}p_ip_j$ is the mean fitness of the population.

The derivation of (17) requires that the simultaneous action of selectional and mutational forces in a small time intervall Δt is of order $(\Delta t)^2$ (compare Akin, 1979). Hadeler (1981) used instead of (17)

$$\dot{p}_i = p_i(m_i - \bar{m}) + \sum_{j=1}^{n}(u_{ij}m_jp_j - u_{ji}m_ip_i). \tag{18}$$

Eq. (18) has the advantage of having the same equilibria as the corresponding discrete time equation, which is biologically less problematic to derive. (17) can be obtained as a limiting case of (18). They have very similar dynamics. For the detailed relations between these equations see Hofbauer (1985). If one realizes that the Malthusian parameter $m_{ij} = b_{ij} - d_{ij}$, where b_{ij} denotes the birth rate of the genotype A_iA_j and d_{ij} its death rate and acknowledges that mutation is only of importance for individuals that are born then the differential equations should read

$$\dot{p}_i = p_i(m_i - \bar{m}) + \sum_{j=1}^{n}(u_{ij}b_jp_j - u_{ji}b_ip_i). \tag{19}$$

All three equations describe dynamical systems on the probability simplex

$$S_n = \{(p_1, \ldots, p_n) \in I\!\!R^n : p_i \geq 0 \text{ and } \sum_{i=1}^{n} p_i = 1\}. \tag{20}$$

For the problems one encounters when modelling continuous time equations the reader is referred to Nagylaki and Crow (1974).

The dynamics of the pure selection equation is well understood. Fisher's Fundamental Theorem of Natural Selection holds, saying that mean fitness is monotone increasing. The pure selection equation is even a gradient with respect to a non-Euclidean metric, the so-called Shahshahani metric (see Hofbauer and Sigmund, 1988). It can be concluded that all trajectories tend to an equilibrium. This is highly non-trivial if continua of solutions exist (Lyubich et al., 1980; Akin and Hofbauer, 1982; Losert and Akin, 1983). It is also well known that if an isolated, stable, interior equilbrium exists it is necessarily globally asymptotically stable with respect to the interior of the simplex.

The situation is far more complicated if mutation is included. Akin (1979) proved a general theorem implying that for any mutation matrix (u_{ij}) that does not satisfy (21) below a selection matrix (m_{ij}) exists such that periodic orbits occur in (17). Hofbauer (1985) proved that stable limit cycles may exist. As an example he presented a model with 3 alleles where all homozygotes have the same fitness and also all heterozygotes. The mutation rates were assumed to be cyclically symmetric.

On the other hand Hadeler (1981) proved that if a model without mutation exhibits an exponentially stable polymorphism then the combined equations (18) have exactly one stationary solution which is also locally exponentially stable, if all mutation rates are identical, i.e. $u_{ij} = \mu$ for all i and j. For three alleles he proved global stability. O'Brien (1985) solved the differential equation (17) for arbitrary mutation rates under the assumption of additive fitness, i.e. $m_{ij} = m_i + m_j$. This, however, reduces (17) to the haploid equation (12). Hence, his result does not go beyond that of Thompson and McBride (1974).

Hofbauer (1985) extended Hadeler's theorem to the case, where the mutation rates satisfy

$$u_{ij} = u_i \text{ for } i \neq j. \tag{21}$$

Since $\sum_{i=1}^{n} u_{ij} = 1$ it follows that $u_{ii} = 1 + u_i - u$ with $u = \sum_{j=1}^{n}$. This assumption on the mutation rate was already used by Kingman (1977) and is known as the house-of-cards model (compare eq. (15)). The following can be proved

Theorem 2.   (Hofbauer, 1985)   *Assume that the mutation rates satisfy* (21) *and* $u \leq 1$. *Then*

$$V(x) = \tfrac{1}{2}\bar{m} + \sum_{i=1}^{n} u_i \log p_i \tag{22}$$

is a Lyapunov function for (17) *and* (17) *is a Shahshahani gradient and all orbits converge to the set of fixed points. If, additionally, the model without mutation admits a stable polymorphism then the mutation-selection equation* (17) *has exactly one interior stationary solution. This solution is even globally stable.*

For (18) analogous results hold, but similar results do not hold for more general mutation rates. Also if selection alone produces a globally stable equilibrium on the boundary of the simplex, this conclusion does not hold, since if mutation is added to a model with two alleles where fitnesses of the genotypes $A_1 A_1$, $A_1 A_2$, $A_2 A_2$ are 1, $1 - s$, $1 - s$ two stable and one unstable equilibrium exists, whereas in the pure selection model the equilibrium with freq(A_1) = 1 is globally stable (Bürger, 1983). Recently, some progress has been achieved in diallelic multi-locus models with equal forward and backward mutation rates at all loci and symmetric viabilities (Barton, 1986; Bürger, 1988c).

4. Mixed strategies and mutations

In this section I consider games with two pure strategies R_1 and R_2 and n mixed strategies $E_1, \ldots, E_n$. Readers who do not like the terminology of evolutionary game theory may think of phenotypes instead of strategies and of frequency dependent selection. Each mixed strategy E_i will be characterized by its probability a_i of using R_1. Without loss of generality $0 \leq a_1 < \ldots < a_n \leq 1$ is assumed. The relative frequency of individuals with phenotype E_i is denoted by p_i. Hence each possible population state $p = (p_1, \ldots, p_n)$ is an element of the simplex S_n (20). Let $\bar{a} = \sum_{i=1}^{n} a_i p_i$ denote the average frequency of R_1 in the population. Throughout, it is assumed that the payoffs A_1 and A_2 for R_1 and R_2 depend only on the population average $\bar{a}$. This is of course a restriction but it is satisfied in many popular games. For example, if the payoffs depend only on (one or repeated) pairwise encounters (as it is the case in the hawk–dove game) $A_1(\bar{a})$ and $A_2(\bar{a})$ are even linear in $\bar{a}$. Subsequently, the function

$$F(\bar{a}) = A_1(\bar{a}) - A_2(\bar{a}) \tag{23}$$

will be considered. Following Taylor and Jonker (1978) and Sigmund (1987) the dynamics is given by

$$\dot{p}_i = p_i(a_i - \bar{a})F(\bar{a}). \tag{24}$$

The dynamics of such games has been investigated recently by Sigmund (1987), who proved that (24) admits a constant of motion and, additionally, is a Shahshahani gradient. The evolution of $\bar{a} \in [a_1, a_n]$ is given by

$$\dot{\bar{a}} = F(\bar{a}) \sum a_i p_i(a_i - \bar{a}) = F(\bar{a}) \text{Var } a. \tag{25}$$

The set $\{p \in S_n : F(\bar{a}) = 0\}$ consists of linear manifolds and is the set of stationary states of (24). In the terminology of Thomas (1985) this is an evolutionarily stable set.

Now I will introduce mutation to the above model. Let u_{ij} designate the mutation rate from E_j to E_i. It is assumed that $u_{ii} > 0$ and $\sum_{j=1}^{n} u_{ji} = 1$ for all i. Then the dynamics is given by

$$\dot{p}_i = p_i(a_i - \bar{a})F(\bar{a}) + \sum_{j=1}^{n} u_{ij}p_j - p_i. \tag{26}$$

If $F(\bar{a}) \equiv 1$ when $a_1 \leq \bar{a} \leq a_n$, (26) reduces to the haploid mutation-selection model (12). It is the purpose of this section to prove the following

Theorem 3. (i) *Suppose that* $a^* \in (a_1, a_n)$ *is a unique equilibrium of (25) such that* $F(a^*) = 0$ *and such that* F *is strictly monotone decreasing on* $[a_1, a_n]$ *(hence* a^* *is globally stable). If the mutation matrix* $U = (u_{ij})$ *is irreducible and if the following convexity property is satisfied*

$$F(a)^2 + (a_i - a)^2(F'(a)^2 - F(a)F''(a)) - (2F'(a) - (a_i - a)F''(a)) \geq 0 \tag{27}$$

$\forall i$ *and* $\forall a \in (\alpha, \beta)$, α, β *being specified below, then (26) admits a unique equilibrium solution* $\hat{p} = (\hat{p}_1, \ldots, \hat{p}_n)$. *Moreover,* $\hat{p}_i > 0$ *holds* $\forall i$.

(ii) *If, additionally, the* u_{ij}*'s are as in the house-of-cards model, i.e. they satisfy (21) and* $u = \sum_{j=1}^{n} u_j \leq 1$, *then (26) is a Shahshahani gradient and the equilibrium* $\hat{p}$ *is globally stable.*

Proof. (i) Designating $f(a, i) = 1 - (a_i - a)F(a)$, the following family of matrices $T(a)$ is introduced

$$[T(a)]_{ij} = \frac{u_{ij}}{f(a, i)}, \quad a \in (\alpha, \beta), \quad i, j = 1, \ldots, n. \tag{28}$$

Here (α, β) is the uniquely determined intervall satisfying $a_1 \leq \alpha < a^* < \beta \leq a_n$ such that $\min_i f(a, i) > 0$ if $a \in (\alpha, \beta)$ and $\min_i f(a, i) < 0$ if $a \in [a_1, a_n] \setminus [\alpha, \beta]$. Its existence follows from the assumptions of Theorem 3(i) upon noticing that

$$\min_i f(a, i) = \begin{cases} 1 - (a_1 - a)F(a), & a > a^*; \\ 1 - (a_n - a)F(a), & a < a^*. \end{cases}$$

It follows that for $a \in (\alpha, \beta)$ each $T(a)$ is a nonnegative irreducible matrix and hence the Perron-Frobenius Theorem applies. Therefore, the spectral radius $r(a)$ of $T(a)$ is an algebraically simple eigenvalue with a strictly positive eigenvector and $T(a)$ has no further positive eigenvectors. We need the following lemmas.

Lemma 1. $\hat{p} = (\hat{p}_1, \ldots, \hat{p}_n)$ *is a stationary point of (26) if and only if* $T(\hat{a})\hat{p} = \hat{p}$ *with* $\hat{a} = \sum a_i \hat{p}_i \in (\alpha, \beta)$ *and* $\hat{p}_i > 0$ *for all* i.

Proof. First observe that the stationary points of (26) are exactly the solutions $\hat{p}$ of

$$\sum_{j=1}^{n} u_{ij} p_j = p_i (1 - (a_i - \bar{a}) F(\bar{a})), \quad \bar{a} = \sum a_i p_i. \tag{29}$$

The ($\Leftarrow$) part of Lemma 1 follows immediately from (28) and (29), since $f(\hat{a}, i) > 0$. To prove the ($\Rightarrow$) part suppose that $\hat{p} = (\hat{p}_1, \ldots, \hat{p}_n)$ satisfies (29). It is sufficient to show that this implies $\hat{p}_i > 0$ and $f(\hat{a}, i) > 0$ for all i. Since U is irreducible, for any $p \in S_n$ with $J(p) = \{i : p_i = 0\} \neq \emptyset$

$$\operatorname{card} J((U + I)p) < \operatorname{card} J(p) \tag{30}$$

holds. Suppose that $J(\hat{p}) \neq \emptyset$. Then $((U + I)\hat{p})_i = \hat{p}_i (2 - (a_i - \hat{a}) F(\hat{a})) = 0$ holds for all $i \in J(\hat{p})$. This contradicts (30) and hence $\hat{p}_i > 0$ holds $\forall i$. Next suppose that $\hat{p}_i > 0 \; \forall i$ but there is an i such that $f(\hat{a}, i) \leq 0$. Then, by (29), $(Up)_i \leq 0$, which is in contradiction to the positivity and stochasticity of U.

Lemma 2. (i) *If* (27) *holds then the function* $a \mapsto \log r(a)$, *where* $r(a)$ *is the spectral radius of the matrix* $T(a)$, *is strictly convex.*

(ii) $a \mapsto r(a)$ *is continuous on* (α, β).

(iii) $r(a^*) = 1$, $\lim_{a \downarrow \alpha} r(a) > 1$ *and* $\lim_{a \uparrow \beta} r(a) > 1$.

Proof. (i) A straightforward calculation shows that (27) implies strict convexity of $a \mapsto \log \dfrac{u_{ij}}{f(a, i)} \; \forall i, j$. By a theorem of Kingman (1961b) this yields the assertion.

(ii) is a consequence of a result of Newburgh (1951).

(iii) Since $T(a^*) = U$ and since $\sum_{j=1}^{n} u_{ji} = 1 \; \forall i$ we have $r(a^*) = 1$. To prove $\lim_{a \downarrow \alpha} r(a) = 1$ we consider two cases. Assume first that $\min_i f(\alpha, i) = 0$. Then $r(a) \geq \max_i \dfrac{u_{ii}}{f(a, i)} \to \infty$ as $a \downarrow \alpha$. If $\min_i f(a, i) > 0$ for all $a \in [\alpha, a^*]$ (and hence $\alpha = a_1$) it follows that $[T(\alpha)]_{ij} = \dfrac{u_{ij}}{f(a_1, i)} \geq u_{ij} \; \forall j$ since $a_i \geq a_1$ and $F(a_1) > 0$. Hence $r(\alpha) \geq 1$ and in fact $r(\alpha) > 1$ because of the convexity property. The third assertion can be shown analogously.

Proof of Theorem 3, continued. To finish part (i) it suffices, due to Lemma 1, to show the existence of a unique $\hat{a} \in (\alpha, \beta)$ such that $r(\hat{a}) = 1$, where the corresponding eigenvector $\hat{p}$ of $T(\hat{a})$ satisfies $\hat{a} = \sum a_i \hat{p}_i$. It follows from Lemma 2 that at least one a exists with $r(a) = 1$, namely $a = a^*$ and at most two. The, possibly, second one is denoted by $\hat{a}$. We consider two cases. First suppose that $a^* = \sum a_i p_i^*$ holds, where p^* is given by $T(a^*)p^* = Up^* = p^*$ (note that U^t is stochastic). Then, using Lemma 2, it can be shown that $r(a) > 1 \; \forall a \in [\alpha, \beta], a \neq a^*$. Therefore, p^* is the desired unique equilibrium solution of (26). If $a^* \neq \sum a_i p_i^*$ then there is an a near a^* such that $r(a) < 1$ and therefore, by Lemma 2, a unique $\hat{a} \neq a^*$ exists with $r(\hat{a}) = 1$. The corresponding eigenvector $\hat{p}$ of $T(\hat{a})$ satisfies (29). Summation over all i yields $\hat{a} = \sum a_i \hat{p}_i$, since $F(\hat{a}) \neq 0$. Hence $\hat{p}$ is the desired unique equilibrium solution of (26). This finishes the proof Theorem 3(i).

(ii) Combining the methods of Hofbauer (1985) and Sigmund (1987) eq. (26) can be written as a replicator equation

$$\dot{p}_i = p_i(g_i - \bar{g}), \tag{31}$$

where $g_i = a_i F(\bar{a}) + \dfrac{u_i}{p_i}$, $\bar{g} = \sum g_i p_i = \bar{a} F(\bar{a}) + u$ and $\bar{a} = \sum a_i p_i$. Following again Hofbauer (1985) and Sigmund (1987), eq. (31) is a Shahshahani gradient with potential

$$V(p) = \int_0^a F(a)\, da + \sum_{i=1}^n u_i \log p_i. \tag{32}$$

In particular, V is a Lyapunov function satisfying $\dot{V}(p) = \sum p_i(g_i - \bar{g})^2 \geq 0$. As a consequence all orbits converge to the set of fixed points. Due to (i) there is a unique fixed point which is, henceforth, globally stable.

As already mentioned, eq. (26) reduces to the haploid mutation-selection equation (12) if $F(a) \equiv 1$. The following two examples illustrate that Theorem 3 is much more generally applicable.

(i) Suppose that F is a linear function (e.g. in the hawk-dove game). Then a^*, $F(a^*) = 0$ is a stable equilibrium of (25) if and only if $F'(a^*) = F'(a) < 0 \; \forall a \in [a_1, a_n]$. Since F is linear, $F'' = 0$ and therefore the convexity condition (27) is satisfied and Theorem 3 can be applied.

(ii) Consider the sex ratio game (e.g. Sigmund, 1987). Then $F(a) = \dfrac{1}{a} - \dfrac{1}{1-a}$. Observe that F is monotone decreasing on (a_1, a_n). Some straightforward but lengthy calculations reveal that F satisfies the convexity condition (27). Therefore Theorem 3 can again be applied.

I do not know whether part (ii) of the above theorem can be extended to more general mutation rates. If this is possible at all, new methods will be required.

Acknowledgments. I thank J. Hofbauer for comments on the manuscript. This work was partially supported by the Austrian *Fonds zur Förderung der wissenschaftlichen Forschung*, Project P5994. Part of it was done during a stay at the International Institute for Applied Systems Analysis, IIASA, Laxenburg, Austria.

References

Akin, E. 1979. The Geometry of Population Genetics. Lect. Notes Biomath. 31. Berlin-Heidelberg-New York. Springer Verlag.

Akin, E., Hofbauer, J. 1982. Recurrence of the unfit. Math. Biosci. 61, 51-63.

Barton, N. 1986. The maintenance of polygenic variation through a balance between mutation and stabilizing selection. Genet. Res. 47, 209-216.

Bulmer, M.G. 1971. Protein polymorphism. Nature 234, 410-411.

Bürger, R. 1983. Dynamics of the classical genetic model for the evolution of dominance. Math. Biosci. 67, 269-280.

Bürger, R. 1986. On the maintenance of genetic variation: Global analysis of Kimura's continuum-of-alleles model. J. Math. Biol. 24, 341-351.

Bürger, R. 1988a. Perturbations of positive semigroups and applications to population genetics. Math. Z. 197, 259-272.

Bürger, R. 1988b. Mutation-selection balance and continuum-of-alleles models. Math. Biosci. To appear.

Bürger, R. 1988c. Linkage and the maintenance of heritable variation by mutation-selection balance. Submitted.

Bürger, R., Wagner, G., Stettinger, F. 1988. How much heritable variation can be maintained in finite populations by a mutation-selection balance? Submitted.

Crow, J.F., Kimura, M. 1964. The theory of genetic loads. Proc. XI Int. Congr. Genet. pp. 495-505. Oxford: Pergamon Press.

Crow, J.F., Kimura, M. 1970. An Introduction to Population Genetics. New York: Harper and Row.

Eigen, M. 1971. Selforganization of matter and the evolution of biological macromolecules. Die Naturwissenschaften 58, 465-523.

Fleming, W.H. 1979. Equilibrium distributions of continuous polygenic traits. SIAM J. Appl. Math. 36, 148-168.

Hadeler, K.P. 1981. Stable polymorphisms in a selection model with mutation, SIAM J. Appl. Math. 41, 1-7.

Hofbauer, J. 1985. The selection mutation equation. J. Math. Biol. 23, 41-53.

Hofbauer, J., Sigmund, K. (1988). Dynamical Systems and the Theory of Evolution. Cambridge Univ. Press. In press.

Kimura, M. 1965. A stochastic model concerning the maintenance of genetic variability in quantitative characters. Proc. Natl. Acad. Sci. USA 54, 731-736.

Kingman, J.F.C. 1961a. On an inequality in partial averages. Quart. J. Math. 12, 78-80.

Kingman, J.F.C. 1961b. A convexity property of positive matrices. Quart. J. Math. 12, 283-284.

Kingman, J.F.C. 1977. On the properties of bilinear models for the balance between genetic mutation and selection. Math. Proc. Camb. Phil. Soc. 81, 443-453.

Kingman, J.F.C. 1978. A simple model for the balance between selection and mutation. J. Appl. Prob. 15, 1-12.

Losert, V., Akin, E. 1983. Dynamics of games and genes: discrete versus continuous time. J. Math. Biol. 17, 241-251.

Lyubich, Yu.I., Maistrovskii, G.D., Ol'klovski, Yu.G. 1980. Selection-induced convergence to equilibrium in a single-locus autosomal population. Problemy Peredachi Informatsii 16, 93-104 (engl. transl.).

Moran, P.A.P. 1976. Global stability of genetic systems governed by mutation and selection. Math. Proc. Camb. Phil. Soc. 80, 331-336.

Moran, P.A.P. 1977. Global stability of genetic systems governed by mutation and selection. II. Math. Proc. Camb. Phil. Soc. 81, 435-441.

Mulholland, H.P., Smith, C.A.B. 1959. An inequality arising in genetic theory. Amer. Math. Monthly 66, 673-683.

Nagylaki, T. 1984. Selection on a quantitative character. In: Human Population Genetics: The Pittsburgh Symposium. (A. Chakravarti, Ed.) New York: Van Nostrand.

Nagylaki, T., Crow, J.F. 1974. Continuous selective models. Theor. Pop. Biol. 5, 257-283.

Newburgh, J.D. 1951. The variation of spectra. Duke Math. J. 18, 165-176.

O'Brien, P. 1985. A genetic model with mutation and selection. Math. Biosci. 73, 239-251.

Ohta, T., Kimura, M. 1973. A model of mutation appropriate to estimate the number of electrophoretically detectable alleles in a finite population. Gen. Res. 22, 201-204.

Ohta, T., Kimura, M. 1975. Theoretical analysis of electrophoretically detectable polymorphisms: Models of very slightly deleterious mutations. Amer. Natur. 109, 137-145.

Scheuer, P., Mandel, S. 1959. An inequality in population genetics. Heredity 13, 519-524.

Sigmund, K. 1987. Game dynamics, mixed strategies, and gradient systems. Theor. Pop. Biol. 32, 114-126.

Taylor, P., Jonker, L. 1978. Evolutionarily stable strategies and game dynamics. Math. Biosci. 40, 145-156.

Thomas, B. Evolutionarily stable sets and mixed strategist models. Theor. Pop. Biol. 28, 332-341.

Thompson, C.J., McBride, J.L. 1974. On Eigen's theory of self-organization of matter and the evolution of biological macromolecules. Math. Biosci. 21, 127-142.

Turelli, M. 1984. Heritable genetic variation via mutation-selection balance: Lerch's zeta meets the abdominal bristle. Theor. Pop. Biol. 25, 138-193.

Turelli, M. 1986. Gaussian versus non-Gaussian genetic analyses of polygenic mutation-selection balance. pp. 607-628. In: Evolutionary Processes and Theory (edt. by S. Karlin and E. Nevo). New York: Academic Press.

Acta Applicandae Mathematicae **14** (1989), 91–102.
© 1989 *by IIASA*.

Pair Formation in Age-Structured Populations

K.P. Hadeler

Universität Tübingen
Lehrstuhl für Biomathematik
Auf der Morgenstelle 10
D-7400 Tübingen, West Germany

AMS Subject Classification (1980): 92A15
Key words: age structure, marriage function, quasimonotone systems

Introduction. In population theory birth and death can be modeled, to a certain extent, by linear equations, but the formation of pairs is a nonlinear phenomenon. Separation of pairs can again be described by linear equations. In particular in human demography it is a long-standing problem by what type of equation pair formation shall be described, in other words, to find an appropriate marriage function. From the work of Kendall (1949), Keyfitz (1972), McFarland (1972), Parlett (1972), Pollard (1973), Fredrickson (1971), Staroverov (1977), Pollak (1987) it is quite clear that mass action kinetics is not appropriate, the marriage law must be homogeneous of degree one. Several special laws have been proposed, e.g. the harmonic mean and the minimum law, but there seems to be no law which is rigorously derived from a microscopic description of the pair formation process. On the other hand ist is obvious what the general properties of a marriage function should be (Fredrickson 1971, see properties 1), 2), 3) below). For such functions we derive a theory of pair formation, that continues and in some sense completes earlier work on this topic. In (Hadeler et al. 1988) we have developed an approach to homogeneous evolution equations, and this theory provides the appropriate framework for pair formation models. In fact we have given a complete analysis of the existence conditions for equilibria and of the global stability problem for these models, for continuous time and in the absence of age structure.

Apart from the immediate application to human demography pair formation models for bisexual populations are necessary prerequisites for the modeling of the spread of sexually transmitted diseases (Dietz and Hadeler 1987). Also there is experimental work on pair formation in animals, in particular in Drosophila and also theoretical research related to such experiments (B.Wallace 1985,1987, Vasco and Richardson 1985). General problems of sex and evolution have been discussed by Williams (1975) and by Karlin and Lessard (1986).

In the models which contain only three one-dimensional variables for singles of both sexes and pairs it is not quite clear whether the newly recruited individuals should be interpreted as newborns or as those who enter the sexually active phase of their life. Therefore a more realistic description should comprise age structure. Mating models for age structured populations have been proposed by most of the authors cited above, in particular Staroverov (1977) has formulated very general models for continuous time, however without much analysis. Keyfitz (1972) has applied such models to demographic data, for several choices of marriage functions.

This work has been supported by the Deutsche Forschungsgemeinschaft

In the present note we discuss the modeling problem, formulate a class of models, and derive some results on persistent distributions.

1.Models for pair formation

The state of the population is defined by three variables x, y, p which describe the densities of female and male singles, and of pairs of a female and a male partner. Then $x + p$, $y + p$ are the densities of all females and of all males, respectively, and $x + y + 2p$ is the total population density.

The parameters of the model are the following. Let κ_x, κ_y be the birth rates of males and females, and let μ_x, μ_y be the corresponding death rates. Let σ be the rate of separation of pairs. These constants are all assumed positive. The processes of birth, death, and separation are assumed to be linear. The process of pair formation is essentially nonlinear. It is governed by an interaction function φ of two variables, $\varphi : \mathbb{R}_+^2 \to \mathbb{R}_+$, with the following properties,

1. Preservation of positivity:

$$\varphi(x, 0) = \varphi(0, y) = 0 \quad \text{for all} \quad x, y \geq 0.$$

2. Homogeneity or scale invariance:

$$\varphi(\alpha x, \alpha y) = \alpha \varphi(x, y) \quad \text{for all} \quad \alpha \geq 0.$$

3. Monotonicity:

$$u \geq 0, \ v \geq 0 \Rightarrow \varphi(x + u, y + v) \geq \varphi(x, y).$$

Furthermore φ is (at least piecewise) continuously differentiable on $\mathbb{R}_+^2 \setminus \{0\}$. An important class of functions φ are derived from the means

$$\varphi(x, y) = \rho(\beta x^\alpha + (1 - \beta) y^\alpha)^{1/\alpha} \tag{1.1}$$

where $0 < \beta < 1$ and $\alpha \in [-\infty, 0]$, and $\rho > 0$.
For $\alpha = -1$ we obtain the harmonic mean

$$\varphi(x, y) = \rho \frac{xy}{\beta x + (1 - \beta)y} , \tag{1.2}$$

in particular, for $\beta = \frac{1}{2}$,

$$\varphi(x, y) = 2\rho \frac{xy}{x + y}. \tag{1.3}$$

We obtain the geometric mean

$$\varphi(x, y) = \rho x^\beta y^{1-\beta} \tag{1.4}$$

in the limit $\alpha \to 0$ and the minimum function

$$\varphi(x, y) = \rho min(x, y) \tag{1.5}$$

for $\alpha \to -\infty$.
With these assumptions the model for a bisexual population is given by the following differential equations

$$\begin{aligned}
\dot{x} &= (\kappa_x + \mu_y + \sigma)p - \mu_x x - \varphi(x, y), \\
\dot{y} &= (\kappa_y + \mu_x + \sigma)p - \mu_y y - \varphi(x, y), \\
\dot{p} &= -(\mu_x + \mu_y + \sigma)p + \varphi(x, y).
\end{aligned} \tag{1.6}$$

If one assumes that recruitment occurs on a slower time scale, i.e. that the newborns are produced at a constant rate unrelated to the number of pairs, then the equations are replaced by

$$\begin{aligned}
\dot{x} &= \kappa_x + (\mu_y + \sigma)p - \mu_x x - \varphi(x, y), \\
\dot{y} &= \kappa_y + (\mu_x + \sigma)p - \mu_y y - \varphi(x, y), \\
\dot{p} &= -(\mu_x + \mu_y + \sigma)p + \varphi(x, y).
\end{aligned} \tag{1.7}$$

These equations have been used in the model for sexually transmitted diseases in (Dietz and Hadeler 1987).

In a very simplified model the birth and death process plays no role, and the equations become

$$\begin{aligned}
\dot{x} &= \sigma p - \varphi(x, y), \\
\dot{y} &= \sigma p - \varphi(x, y), \\
\dot{p} &= -\sigma p + \varphi(x, y).
\end{aligned} \tag{1.8}$$

The system (1.8) is equivalent to a single differential equation

$$\dot{p} = -\sigma p + \varphi(\bar{x} - p, \bar{y} - p)$$

where $\bar{x}$, $\bar{y}$ are the total densities of females and males, respectively.

The general system (1.6) is much more difficult to analyze. In view of the homogeneity property stationary solutions cannot be expected. Instead one looks for persistent distributions, i.e. for exponential solutions of the form

$$(x, y, p) = (\bar{x}, \bar{y}, \bar{p})e^{\lambda t}. \tag{1.9}$$

The existence and stability of exponential solutions can be discussed in several ways (for a systematic approach to homogeneous evolution equations see Hadeler et al. 1987, 1988). In the present case one can reduce the problem to a two-dimensional system (to which the theorem of Poincaré and Bendixson applies) by introducing either barycentric coordinates

$$u = \frac{x}{x+y+p}, \quad v = \frac{y}{x+y+p}, \quad w = \frac{p}{x+y+p}, \tag{1.10}$$

or "projective" variables

$$\xi = \frac{x}{p}, \quad \eta = \frac{y}{p}. \tag{1.11}$$

The first transformation leads to the system

$$
\begin{aligned}
\dot{u} &= (\kappa_x + \mu_y + \sigma)w - \mu_x u - \varphi(u,v) - \phi(u,v,w)u, \\
\dot{v} &= (\kappa_y + \mu_x + \sigma)w - \mu_y v - \varphi(u,v) - \phi(u,v,w)v, \\
\dot{w} &= -(\mu_x + \mu_y + \sigma)w + \varphi(u,v) - \phi(u,v,w)w,
\end{aligned}
\tag{1.12}
$$

$$\phi = (\kappa_x + \kappa_y + \sigma)w - \mu_x u - \mu_y v - \varphi(u,v)$$

on the triangle $S = \{u,v,w \geq 0, \quad u + v + w = 1\}$, the second to

$$
\begin{aligned}
\dot{\xi} &= \kappa_x + (\mu_y + \sigma)(1 + \xi) - (1 + \xi)\varphi(\xi,\eta), \\
\dot{\eta} &= \kappa_y + (\mu_x + \sigma)(1 + \eta) - (1 + \eta)\varphi(\xi,\eta)
\end{aligned}
\tag{1.13}
$$

on $\mathbb{R}^2_+$. The system (1.13) is quasimonotone (or Kamke monotone, see Hirsch 1982), every bounded trajectory converges to a stationary point. The system (1.12) is rather close to so-called replicator equations (see Hofbauer and Sigmund 1984). In (Hadeler et al. 1988) we have given a complete description of the global behavior of these systems. We keep all parameters fixed except the mortalities μ_x, μ_y. Depending on the relative size of the parameters μ_x, μ_y there are three qualitatively different situations (see figures 1 and 2).

Theorem: *If*

$$\mu_x < \mu_y - \frac{\kappa_y \varphi_y(1,0)}{\mu_y + \sigma + \varphi_y(1,0)}$$

then the positive stationary state does not exist, and $(1,0,0)$ *is stable,* $(0,1,0)$ *is unstable.*
If

$$\mu_y > \mu_x - \frac{\kappa_x \varphi_x(0,1)}{\mu_x + \sigma + \varphi_x(0,1)}$$

and

$$\mu_x > \mu_y - \frac{\kappa_y \varphi_y(1,0)}{\mu_y + \sigma + \varphi_y(1,0)}$$

then the positive stationary solution exists, is stable and $(1,0,0)$, $(0,1,0)$ *are unstable.*
If

$$\mu_y < \mu_x - \frac{\kappa_x \varphi_x(0,1)}{\mu_x + \sigma + \varphi_x(0,1)}$$

then the positive stationary state does not exist, the solution $(1,0,0,)$ *is unstable, but* $(0,1,0)$ *is stable.*

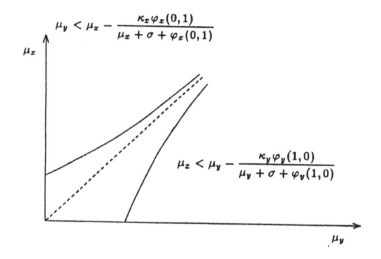

Figure 1: The μ_x, μ_y parameter plane

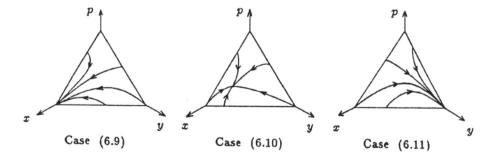

Figure 2: Phase portraits of the pair formation model

Because periodic solutions can be excluded by studying (1.13), local stability implies in this case global stability.

The theorem also clarifies the dynamics of the original system (1.6). Stationary points of (1.12) correspond to exponential solutions of (1.6).

2.Pair formation and age structure

In this section we address the problem of pair formation in a population with age structure. As mentioned in the introduction such models have been proposed by several authors. It appears that a thorough investigation of these models has not been performed. Perhaps this is due to the fact that models for pair formation are indeed difficult to analyze. The finite-dimensional case is already complicated. Therefore, one should restrict attention to the simplest possible situation.

Again, the variables x, y, p denote females, males, and pairs, respectively. It is convenient to denote age of females and males by different variables. Let $x = x(a)$ be the density of females of age a, i.e. the number of females with age between a_1 and a_2 is

$$\int_{a_1}^{a_2} x(a)da.$$

Similarly, $y(b)$ is the density of males of age b. Pairs are classified according to the ages of the female and male partners. Staroverov has observed that one can also keep track of the duration of the "marriage". A pair is of type (a, b, c) if the female has age a, the male has age b, and the pair has age c. The function $p(a, b, c)$ is the density of pairs. Hence

$$\int_{a_1}^{a_2} \int_{b_1}^{b_2} \int_{c_1}^{c_2} p(a, b, c)dcdbda$$

is the number of pairs, where the female has age between a_1 and a_2, the male has age between b_1 and b_2, and the pair exists already during a time period of length c, where $c_1 \leq c \leq c_2$. Of course $p(a, b, c) \equiv 0$ for $c < min(a, b)$.

For simplicity we assume that mortality is independent of the "marital" status. Let $\mu_x(a)$, $\mu_y(b)$ be the age-dependent mortalities of females and males, respectively. Let $\beta_x(a, b, c)$, $\beta_y(a, b, c)$ be the number of offspring of female or male sex, respectively, produced by a pair of type (a, b, c). Let $\sigma(a, b, c)$ be the separation rate of pairs of type (a, b, c). In order to model the act of pair formation, one can introduce rather complicated nonlinear functions (see e.g. Staroverov 1977).

A plausible mating function for an age structured population can be derived using the concept of preference distributions. Assume a female of age a considers males of age in $[b, b + \Delta b]$ as possible partners with probability $g(a, b)\Delta b$. Then we call $g(a, b)$ the preference distribution of a female of age a. Of course

$$g(a, b) \geq 0, \quad \int_0^\infty g(a, b)db = 1.$$

Similarly let $h(a, b)$ be the preference distribution of a male of age b,

$$h(a, b) \geq 0, \quad \int_0^\infty h(a, b)da = 1.$$

Let $x = x(a)$, $y = y(b)$ be the (non-normalized) densities of females and males. If these individuals are brought into contact according to preference, then in the joint age class $[a, a + \Delta a) \times [b, b + \Delta b)$ we find $x(a)g(a, b)\Delta b \Delta a$ females and $h(a, b)y(b)\Delta a \Delta b$ males. Within this class the ages a of females and b of males are approximately constant. Thus the harmonic mean law is appropriate within this class, whereby the coefficient $\rho = \rho(a, b)$ depends on the ages a, b. Hence the rate of pair formation is

$$\varphi(x, y)(a, b) = 2\rho(a, b)\frac{x(a)g(a, b)h(a, b)y(b)}{x(a)g(a, b) + h(a, b)y(b)}. \tag{2.1}$$

Here φ is a function which maps a pair of densities x, y into a function $\varphi(x, y)$ of two variables a, b. Thus $\varphi(x, y)(a, b)$ is the function $\varphi(x, y)$ evaluated at a, b.
Of course this function is still special in the sense that the preference distribution does not depend on the partners actually available. Notice that the function ρ is not necessarily constant. $\rho(a, b)\Delta t$ is the probability that a female of age a and a male of age b are forming a pair (during Δt) under the condition that other choices are not available.

This function φ has the following properties: $\varphi(x, y)(a, b)$ vanishes if either $x(a)$ or $y(b)$ is zero, $\varphi(x, y)(a, b)$ is nondecreasing in $x(a)$ and in $y(b)$. The function φ is homogeneous of degree 1.
Consider the caricature where females prefer almost exclusively males of the same age, and conversely. Then $\varphi(x, y)(a, b) \approx 0$ for $|a - b|$ large and

$$\varphi(x, y)(a, a) \approx 2\rho(a, a) \cdot \frac{x(a)y(a)}{x(a) + y(a)}. \tag{2.2}$$

If the two sexes occur with the same density in each age class then

$$\varphi(x, x)(a, a) = \rho(a, a)x(a). \tag{2.3}$$

Then one has essentially a one sex-model where $\rho(a, a)$ is the age-dependent pair formation rate.

In another special case there are no preferences but just a pair formation rate. Then

$$\varphi(x, y)(a, b) = 2\rho(a, b)\frac{x(a)y(b)}{x(a) + y(b)}. \tag{2.4}$$

If in each age class females and males come in equal numbers, $x(a) = y(a)$, then

$$\varphi(x, x)(a, b) = 2\rho(a, b)\frac{x(a)x(b)}{x(a) + x(b)} \tag{2.5}$$

which is quite different from (2.2). In a discrete model (2.2) would be a diagonal matrix and (2.5) very different from a diagonal matrix.

Now the complete model consists of three hyperbolic differential equations

$$x_t + x_a + \mu_x(a)x + \int_0^\infty p(t, a, b, 0)db$$
$$- \int_0^\infty \int_0^\infty \mu_y(b)p(t, a, b, c)dbdc - \int_0^\infty \int_0^\infty \sigma(a, b, c)p(t, a, b, c)dbdc = 0,$$

$$y_t + y_b + \mu_y(b)y + \int_0^\infty p(t, a, b, 0)da \tag{2.6}$$
$$- \int_0^\infty \int_0^\infty \mu_x(a)p(t, a, b, c)dadc - \int_0^\infty \int_0^\infty \sigma(a, b, c)p(t, a, b, c)dadc = 0,$$

$$p_t + p_a + p_b + p_c + \mu_x(a)p + \mu_y(b)p + \sigma(a, b, c)p = 0,$$

together with boundary conditions

$$x(t, 0) = \int_0^\infty \int_0^\infty \int_0^\infty \beta_x(a, b, c)p(t, a, b, c)dadbdc,$$
$$y(t, 0) = \int_0^\infty \int_0^\infty \int_0^\infty \beta_y(a, b, c)p(t, a, b, c)dadbdc,$$

$$\tag{2.7}$$

$$p(t, a, 0, c) = 0,$$
$$p(t, 0, b, c) = 0,$$
$$p(t, a, b, 0) = \varphi(x, y)(t, a, b),$$

and initial conditions
$$x(0, a) = x_0(a),$$
$$y(0, b) = y_0(b), \tag{2.8}$$
$$p(0, a, b, c) = p_0(a, b, c).$$

The most interesting solutions of these equations are the persistent distributions

$$x(t, a) = \bar{x}(a)e^{\lambda t}, \quad y(t, b) = \bar{y}(b)e^{\lambda t}, \quad p(t, a, b, c) = \bar{p}(a, b, c)e^{\lambda t}. \tag{2.9}$$

Here we shall not enter the problem of existence of these solutions.

If the separation rate σ and the birth rates β_x, β_y do not depend on the age c of the marriage but only on the ages of the partners, then the equations $(2.6)(2.7)$ can be integrated over c and yield the simpler equations

$$x_t + x_a + \mu_x(a)x + \int_0^\infty \varphi(x,y)(t,a,b)db$$
$$- \int_0^\infty \int_0^\infty (\mu_y(b) + \sigma(a,b))p(t,a,b)db = 0,$$
$$y_t + y_b + \mu_y(b)y + \int_0^\infty \varphi(x,y)(t,a,b)da \qquad (2.10)$$
$$- \int_0^\infty \int_0^\infty (\mu_x(a) + \sigma(a,b))p(t,a,b)da = 0,$$
$$p_t + p_a + p_b + (\mu_x(a) + \mu_y(b) + \sigma(a,b))p - \varphi(x,y)(t,a,b) = 0,$$

$$x(t,0) = \int_0^\infty \int_0^\infty \beta_x(a,b)p(t,a,b)dadb,$$
$$y(t,0) = \int_0^\infty \int_0^\infty \beta_y(a,b)p(t,a,b)dadb, \qquad (2.11)$$
$$p(t,a,0) = p(t,0,b) = 0.$$

In the following we restrict our attention to a very special case where all parameter functions are symmetric with respect to females and males, where in each age class there are equal numbers of females and males, and where pair formation occurs only within the same age class. Then it is sufficient to consider singles x and pairs p only, and the equations become linear,

$$x_t + x_a + \mu(a)x + \rho(a)x - (\mu(a) + \sigma(a))p = 0,$$
$$p_t + p_a + (2\mu(a) + \sigma(a))p - \rho(a)x = 0, \qquad (2.12)$$

$$x(t,0) = \int_0^\infty \beta(a)p(t,a)da,$$
$$p(t,0) = 0, \qquad (2.13)$$

together with the appropriate initial conditons.
Persistent solutions have the form

$$x(t,a) = x(a)e^{\lambda t}, \quad p(t,a) = p(a)e^{\lambda t}$$

where the functions $x(a), p(a)$ satisfy

$$x_a + (\mu(a) + \rho(a) + \lambda)x - (\mu(a) + \sigma(a))p = 0,$$
$$p_a + (2\mu(a) + \sigma(a) + \lambda)p - \rho(a)x = 0, \qquad (2.14)$$

$$x(0) = \int_0^\infty \beta(a)p(a)da,$$

$$p(0) = 0. \tag{2.15}$$

The persistent solutions can be found explicitly. Solutions of the linear system (2.14), for $\lambda = 0$, i.e. of the system

$$\begin{pmatrix} x_a \\ p_a \end{pmatrix} = \begin{pmatrix} -(\mu + \rho) & \mu + \sigma \\ \rho & -(2\mu + \sigma) \end{pmatrix} \begin{pmatrix} x \\ p \end{pmatrix} \tag{2.16}$$

satisfy

$$(x + p)_a = -\mu(a)(x + p).$$

Hence

$$x_a = -(2\mu + \rho + \sigma)x + (\mu + \sigma)(x_0 + p_0)e^{-M(a)},$$
$$p_a = -(2\mu + \rho + \sigma)p + \rho(x_0 + p_0)e^{-M(a)}, \tag{2.17}$$

where

$$M(a) = \int_0^a \mu(s)ds. \tag{2.18}$$

The solutions are

$$x(a) = x_0 e^{-\int_0^a (2\mu+\rho+\sigma)ds} + \int_0^a e^{-\int_s^a (2\mu+\rho+\sigma)d\tau} e^{-M(s)}(\mu + \sigma)ds(x_0 + p_0),$$

$$p(a) = p_0 e^{-\int_0^a (2\mu+\rho+\sigma)ds} + \int_0^a e^{-\int_s^a (2\mu+\rho+\sigma)d\tau} e^{-M(s)}\rho\, ds\, (x_0 + p_0).$$

$$\tag{2.19}$$

The solutions of interest of the system (2.14) are obtained for $p_0 = 0$,

$$x(a) = x_0 \left[e^{-\int_0^a (2\mu+\rho+\sigma)ds} + \int_0^a e^{-\int_s^a (2\mu+\rho+\sigma)d\tau} e^{-M(s)}(\mu + \sigma)ds \right] e^{-\lambda a},$$

$$p(a) = x_0 \int_0^a e^{-\int_s^a (2\mu+\rho+\sigma)d\tau} e^{-M(s)}\rho ds \cdot e^{-\lambda a}. \tag{2.20}$$

Clearly this solution is nonnegative.

From the recruitment law (2.15) it follows, after division by x_0, that λ satisfies the equation

$$1 = \int_0^\infty \beta(a) \int_0^a e^{-\int_s^a (2\mu+\rho+\sigma)d\tau} e^{-M(s)} \rho(s)ds e^{-\lambda a} da. \tag{2.21}$$

This characteristic equation can be written in a more convenient way as

$$\int_0^\infty \beta(a)e^{-M(a)} \int_0^a e^{-\int_s^a (\mu+\sigma)d\tau} e^{-\int_s^a \rho d\tau} \rho(s)ds e^{-\lambda a} da = 1. \tag{2.22}$$

The largest real root of this equation is the exponent of the persistent solutions. This exponent in the analogon of "Lotka's r" in the usual demographic models. The interpretation of equation (2.22) is as follows. The factor $\beta(a)exp[-M(a)]$ is the usual kernel of the Lotka-Sharpe model. The second factor, which for $\mu = \sigma = 0$ would just be $1 - exp(- \int_0^a \rho(s)ds)$ gives the number at pairs still present in each age class.

The exponent λ is increased if in some age classes either μ, σ are decreased or β, ρ are increased. This conclusion is obvious for μ, σ, β. The dependence on ρ is less obvious, since the kernel in (2.22) contains ρ as a factor and in the exponent. However

$$\int_0^a e^{-\int_s^a (\mu+\rho+\sigma)d\tau} \rho(s)ds = 1 - e^{-\int_0^a (\mu+\sigma+\rho)d\tau} - \int_0^a e^{-\int_s^a (\mu+\sigma+\rho)d\tau}(\mu+\sigma)ds$$

and the right hand side is clearly not decreased if ρ is increased in any age class. Hence λ is not decreased if ρ is increased.

In the case where all coefficients are constant, the solutions (2.20) are

$$\begin{pmatrix} x \\ p \end{pmatrix} = \frac{x_0}{\mu+\sigma+\rho} e^{-(\mu+\lambda)a} \begin{pmatrix} (\mu+\sigma) + \rho e^{-(\mu+\sigma+\rho)a} \\ \rho(1 - e^{-(\mu+\sigma+\rho)a}) \end{pmatrix} \qquad (2.23)$$

and the characteristic equation is

$$\frac{\beta\rho}{(\mu+\lambda)(2\mu+\rho+\sigma+\lambda)} = 1. \qquad (2.24)$$

Only the greater solution of this equation is relevant, since the other does not lead to an integrable function. Hence the exponent of the persistent solution is

$$\lambda = \frac{1}{2}[\sqrt{(\mu+\sigma+\rho)^2 + 4\beta\rho} - (3\mu+\sigma+\rho)]. \qquad (2.25)$$

References.

Dietz, K., Hadeler, K.P., Epidemiological models for sexually transmitted diseases. J.Math.Biol. 26, 1-25 (1988)

Dowse, H.B., Ringo, J.M., Barton, K.M., A model describing the kinetics of mating in *Drosophila*, J. Theor. Biol. 121, 173-183 (1986)

Hadeler, K.P., Waldstätter, R., Wörz-Busekros, A., Models for pair formation. In: Conference Report, Deutsch-Französisches Treffen über Evolutionsgleichungen, Blaubeuren, 3.-9. Mai 1987, Semesterbericht Funktionalanalysis Tübingen 1987, p.31-40.

Hadeler, K.P., Waldstätter, R., Wörz-Busekros, A., Models for pair formation in bisexual populations. J.Math.Biol. submitted

Hirsch, M.W., Systems of differential equations which are competitive or cooperative I. Limit sets. SIAM J.Math.Anal.13,167-179(1982)

Hofbauer, J., Sigmund, K., Evolutionstheorie und dynamische Systeme. Paul Parey Verlag Berlin Hamburg 1984

Karlin, S., Lessard, S., Theoretical studies on sex ratio evolution. Monographs in Population Biology Vol.22, Princeton University Press. Princeton New Jersey 1986

Kendall, D.G., Stochastic processes and population growth. Roy.Statist.Soc., Ser B,2,230-264(1949)

Keyfitz, N., The mathematics of sex and marriage. Proc. of the Sixth Berkeley Symposion on Mathematical Statistics and Probability. Vol.IV: Biology and Health,p.89-108(1972)

McFarland, D.D.,Comparison of alternative marriage models. Population Dynamics (T.N.T.Greville, ed.),p.89-106, Academic Press, New York London (1972)

Fredrickson, A.G., A mathematical theory of age structure in sexual populations: Random mating and monogamous marriage models. Math.Biosciences 10, 117-143 (1971)

Parlett, B., Can there be a marriage function? Population Dynamics (T.N.T. Greville, ed.),p.107-135, Academic Press, New York London (1972)

Pollak, R.E., The two-sex problem with persistent unions: a generalization of the birth matrix-mating rule model, Theor. Pop. Biol. 32, 176-187 (1987)

Pollard, J.H., Mathematical models for the growth of human populations, Chapt.7: The two sex problem. Cambridge University Press, Cambridge 1973

Staroverov, O.V., Reproduction of the structure of the population and marriages. (Russian) Ekonomika i matematičeskije metody 13,72-82(1977)

Vasco, D.A., Richardson, R.H., On the theory of sexual reproduction kinetics with comments on some recent experimental results . Behavioral Sciences, 30, 134-149(1985)

Wallace, B., Mating Kinetics in Drosophila. Behavioral Science 30, 149-154 (1985)

Wallace, B., Kinetics of mating in Drosophila. I, *D. melanogaster*, an *ebony* strain, preprint. Dept. Biol., Virginia Polytechnic and State University

Williams, G.C., Sex and evolution. Monographs in Population Biology Vol.8, Princeton University Press, Princeton New Jersey (1975)

N^0 meaning $\{0\}$. (This is the well-known Ulam-Harris construction.)

If now $L, M \subset I$, we may write $M \prec L$, whenever any individual in L stems from some individual in M. (In our mathematical notation, y stems from x, in symbols $x \prec y$, if x constitutes the first part of the vector y). If we want to investigate the plausibility of some assertion about the lives of individuals who belong to L, or stem from L, and know everything about all individuals in M it is obvious that we can do no better with information also about individuals preceding those in M. Thus, the process evolves as a Markov field on the space of subsets of I, partially ordered by descent (denoted by $\prec$).

It can be proved that the process even has the *strong Markov property*: It starts anew not only from fixed sets of individuals but also from those that are random, provided that this randomness only derives from the outcomes of lives of individuals not in the progeny of the set itself. An example of a random set which is *predetermined* (or *optional*) in this sense might be the set of all those born before a certain time, another the set of all children of some given individual who are ever born. In both cases the sets are obviously random, but all randomness arises from individuals preceding those in the set itself or included in it. The strong Markov property is important because it makes it possible to divide the population into, conditionally, independent subpopulations. Then the whole arsenal of methods developed within probability theory for analysis of sums of independent random entities can be called upon.

The transition from a process evolving over partially ordered sets of individuals to a process in real time is easy: Eve was born at time zero. The birth time of any other individual is the sum of all the maternal ages at the bearings in the genealogical line that leads from Eve to the individual in question. Thus, the *general branching process* arises.

But what can be proved for such a general scheme? Astonishingly much and, disregarding technical conditions, which may be mathematically interesting but of little relevance for the basic principles, the results can be sketched as follows: The population either dies out, or grows exponentially. The extinction probability can be determined from the off-spring number distribution, the so called *reproduction law*. (This is a famous fact from simple branching processes.) The exponential growth occurs at the rate of the *Malthusian parameter*, usually denoted α, also determined by the individual reproductive behaviour, and indeed by the kernel $\mu(s, A \times B)$, giving the expected number of children with types in A, borne by an individual of type s, while of age in B. Here, of course $A \subset S$ and B is a subset of the non-negative real line, R_+. In terms of the reproduction process ξ,

$$\mu(s, A \times B) = E_s[\xi(A \times B)],$$

E_s being expectation with respect to $P(s, \cdot)$. This reproduction kernel, multiplied by $e^{-\alpha t}$, has a positive eigenfunction h,

$$h(s) = \int_{S \times R_+} h(r) e^{-\alpha t} \mu(s, dr \times du),$$

and a *stable type distribution* π such that

$$\pi(A) = \int_{A \times R_+} e^{-\alpha t} \mu(s, A \times dt) \pi(ds).$$

Note that π need not be a probability distribution. However, we can and shall choose h so that $\int h d\pi = 1$. When π is finite, it can be normalized, too. Then it has the interpretation of being the limiting type distribution among new-borns.

These entities are crucial in determining a *stable population composition*, which can be shown to emerge asymptotically, as time passes, in any branching population not dying out. At this stage it would require too much notation to give its general form explicitly in terms of the life law. What matters is that special margins of the stable population composition can be calculated, and actually rather easily so, in any case of interest.

A classical example is the famous stable age distribution of one-type populations, known in principle already to Euler: If an individual is sampled at random from an old, or stably composed, population, then its age follows the distribution function

$$A(t) = \frac{\int_0^t e^{-\alpha u}(1 - L(u))du}{\int_0^\infty e^{-\alpha u}(1 - L(u))du},$$

L denoting the *life length distribution* of a new born individual of the one and only type under consideration. In the general multi-type case let $\lambda : \Omega \to R_+$ denote the individual *life span* and write

$$P(\pi, \cdot) = \int_S P(s, \cdot)\pi(ds).$$

Then the stable age-distribution takes the form

$$A(t) = \frac{\int_0^t e^{-\alpha u} P(\pi, \lambda > u)du}{\int_0^\infty e^{-\alpha u} P(\pi, \lambda > u)du}.$$

In general the kernel $P(s, \cdot)$ defines a probability measure $\tilde{P}$, the *stable population measure*, over a quite complex doubly infinite pedigree space, centered around an arbitrary, or typical, individual of the stable population. The interpretation is that this individual is sampled at random from an old, or otherwise stabilized, population and $\tilde{P}$ gives the probabilities of various properties of the sampled individual, including properties of his progeny or ancestors or brothers and sisters. Thus the stable age distribution is but one simple margin of $\tilde{P}$,

$$A(t) = \tilde{P}(age \le t).$$

Other examples are the mitotic index of a cell population in balanced exponential growth - or even in a growth of circadian variation, where types can be used to mirror clock time, cf. Jagers and Nerman, 1985. Indeed, the mitotic index is nothing but the probability of a random individual being in mitosis, which is easily expressible through $\tilde{P}$, provided the individual cell cycle and mitosis spans are known in distribution. A simple example of a relational property, i. e. a property concerning the relation between the random individual and others, is given by the probability of being first-born,

$$\tilde{P}(first\ born) = E_\pi[e^{-\alpha\tau(1)}],$$

E_π being expectation with respect to $P(\pi, \cdot)$.

This illustrates how $\tilde{P}$ is usually easily expressed in terms of the life law. For virtually any application such expressions provide the asked for bridge between population composition and individual properties. With their help we can thus predict population behaviour from knowledge of individual lives, and conversely infer about individual properties from observations on populations.

The exact general form of $\tilde{P}$ is quite intricate. We shall return to it in the next section. For a strict description cf. Jagers and Nerman, 1984. It suffices here to say that if we propose to measure individuals, so that the measure at age t of a type-s-individual leading life ω is $\chi(s, t, \omega)$, then under general conditions, the χ-value of a random individual among all those born will have the expected value $E_\pi[\hat{\chi}(\alpha)]$, hat denoting Laplace-Riemann transform, $\hat{f}(\alpha) = \int_0^\infty e^{-\alpha t} f(t)dt$ and $\chi(t)$ being short for $\chi(\cdot, t, \cdot)$.

Since the indicator function $1_{[0,\lambda)}(t)$ gives a living individual of age t measure one, we recognize in the denominator of the stable age-distribution the proportion of individuals alive, among all those born. And if χ also vanishes as soon as the individual dies (if not earlier) the ratio

$$\frac{E_\pi[\hat{\chi}(\alpha)]}{\int_0^\infty e^{-\alpha u} P(\pi, \lambda > u)du}$$

gives the mean χ-value in a balanced population of living individuals. The reader should note the distinction beteen sampling the random individual from among all presently alive or from among all those born by now. For most applications it is the former concept that is relevant, but mathematically the latter comes first. This is because sampling from subsets can then be expressed through conditional probabilities, like the ratio above, and then the composition of many subpopulations can be investigated as well. As an example the fraction of labelled mitoses from cell kinetics might serve, being the ratio between the number of labelled mitotic cells and the number of all cells in mitosis.

3. Into the Mathematics.

To make the preceding more precise we must equip the life space with a sigma-algebra of subsets, $\mathcal{A}$, and the type space with a (countably generated) sigma-algebra $\mathcal{S}$. Then the *population space*

$$(S \times \Omega^I, \mathcal{S} \times \mathcal{A}^I)$$

can be defined as the outcome space of the whole population process.

On the population space the projections $U_M : S \times \Omega^I \to \Omega^M, M \subset I$, are defined. Particularly important are $S_x = U_{\{y \in I; y \succ x\}}, x \in I$, just considering the progeny of x, and U_x, short for $U_{\{x\}}$, which just looks at x's life. They define further sigma-algebras

$$\mathcal{F}_L = \sigma(U_x; L \not\prec x),$$

which contain all information about the population not stemming from $L \subset I$. Since $L \prec M \Rightarrow \mathcal{F}_L \subset \mathcal{F}_M$, they constitute a filtration.

A basic existence theorem can now be proved by use of Ionesco-Tulcea's theorem, and the fundamental concept of a *line* of individuals (Chauvin, 1986). A line is a set of individuals such that no member stems from any other member of the set. (It is thus, rather, the opposite of a genealogical line.) To state the theorem we need notation for the type of an individual x, ρ_x.

Theorem 1 *There is exactly one probability measure P_s, such that $\rho_0 = s \in S$ and the conditional distribution of U_x given $\mathcal{F}_x = \mathcal{F}_{\{x\}}$ is $P(\rho_x, \cdot)$. For this P_s the Markov property*

$$P_s(S_x \in \cdot \mid \mathcal{F}_x) = P_{\rho_x}$$

holds and if L is a line, then given $\mathcal{F}_L$ the $S_x, x \in L$, are conditionally independent with distribution P_{ρ_x}. (This is the branching property, Chauvin, 1986.)

(The meticulous reader might object that it is actually the pair ρ_x, S_x that has the conditional distribution P_{ρ_x}. But since the first component is degenerate, we allow ourselves this light inexactitude.)

For the corresponding strong Markov theorem, the property of being predetermined, mentioned in the preceding section is needed. It is strictly defined as for the traditional concept of optional or stopping times: A random subset $\mathcal{I}$ of I is *predetermined* if

$$\forall L \subset I : \{\mathcal{I} \prec L\} \in \mathcal{F}_L.$$

Some interesting examples of predetermined sets, defined in terms of the *birth times* σ_x of $x \in I$, are

- $\mathcal{Y}_t = \{x \in I; \sigma_x \leq t\}$, all those born by t,

- $\mathcal{R} = \{x \in I; \sigma_x < \infty\}$, those ever born,

- $\mathcal{Z}_n = N^n \bigcap \mathcal{R}$, the n:th generation,

- $\mathcal{I}_t = \{x \in I; \sigma_{x's mother} < t \leq \sigma_x < \infty\}$.

The last of these, the *coming generation* at t, due to Nerman, 1981, is of particular help in the analysis. Of the four it is only this set and the n:th generation that are lines. But all four are what might be called *covering*, any individual ever born either stems from the set or has some ancestor in it.

Theorem 2 *Let $\mathcal{I}$ be a predetermined line and ϕ_x, $x \in I$, measurable functions $S \times \Omega^I \to [0,1]$. Then, E_s denoting integration with respect to P_s,*

$$E_s[\prod_{x \in \mathcal{I}} \phi_x \circ S_x \mid \mathcal{F}_{\mathcal{I}}] = \prod_{x \in \mathcal{I}} E_{\rho_x}[\phi_x].$$

Here, of course, $\mathcal{F}_{\mathcal{I}}$ denotes the sigma-algebra of events preceding $\mathcal{I}$, defined as usual, $A \in \mathcal{F}_{\mathcal{I}} \Leftrightarrow \forall L \subset I : A \bigcap \{L \prec \mathcal{I}\} \in \mathcal{F}_L$.

The next step of the analysis is to produce an *intrinsic martingale*, which will catch the randomness in population evolution. Recall the eigenfunction h and the Malthusian parameter α, whose precise definition might be the number, assumed positive, such that the kernel $\int_0^\infty e^{-\alpha t} \mu(s, dr \times dt)$ on S yields an operator on the bounded functions on S with spectral radius one. In terms of those we define

$$w_M = \sum_{x \in M} e^{-\alpha \sigma_x} h(\rho_x), \quad x \in M,$$

for any $M \subset I$. It is somewhat misleading to call this a martingale. What is true are statements like

Theorem 3 *If $L \prec M$ are lines, then*

$$E_s[w_M \mid \mathcal{F}_L] \leq w_L.$$

If the maximal generation of elements of M is finite and M is covering, then equality holds.

The *generation* of $x \in I$ of course is its dimension. Thus $\{w_M; M \subset I$ being covering lines of finite maximal generation$\}$ is a martingale. Under the general equivalent of the so called $x \log x$-condition it is even uniformly integrable, and thus has a limit in $L^1(P_s)$, as M filters away. At least this is true for almost all $s \in S$ with respect to the stable type distribution π. In the usual manner for martingales this extends the above theorem to random, predetermined lines. However, at this stage results tend to be terribly technical, and for exact statements and proofs the reader is referred to Jagers, 1988. Let me only state the $x \log x$-condition: Write

$$\tilde{\xi} = \int_{S \times R_+} h(s) e^{-\alpha t} \xi(ds \times dt).$$

Then the condition is that

$$E_\pi[\tilde{\xi} \log \tilde{\xi}] < \infty.$$

By this we end our investigation of model structure and turn to the development of population "size" in real time. As mentioned in the first section "size" could be many things: mass, age, DNA-content, an indicator telling whether the individual at her age now has a property (is in a stage) or not. This versatility is important not only because it yields overall theorems, covering many cases, but also since ratios of different sizes tell about population composition. E. g. if one "size" is the indicator telling whether an individual cell is in mitosis or not and another is the indicator telling if the cell is alive, then their ratios is the mitotic index. The important restriction is that population "size" at any time should be the sum of individual "sizes" at that time. These individual "sizes" might then actually be influenced by factors outside the individual's life, at least by her progeny. In this overview we shall, however, mainly have "sizes" in mind that are functions of the individual's type, age, and own life history.

The basic idea in the analysis is to choose a suitable predetermined covering line, call it $\mathcal{I}$ and then to divide the population like this: The population "size" at $t =$ the "size" of those before $\mathcal{I}$ (little, if $\mathcal{I}$ was cleverly chosen) + the sum of the "sizes" of all the populations stemming from the $x \in \mathcal{I}$, which are now $t-\sigma_x$ time units old and, conditionally given $\mathcal{F}_{\mathcal{I}}$, independent of one another. This combination of recursion and independence solves the problem: One choice of $\mathcal{I}$, the first generation or N, taking expectations and use of Markov renewal theory yields the expected behaviour, as time passes to infinity. Another choice, the coming generations at some time before t, makes it possible to combine classical limit theory for sums of independent random variables with what we know about the intrinsic martingale in order to catch the random variations around the expected development.

To be more precise, we consider a jointly measurable function $\chi : S \times R \times \Omega \to R_+$, a so called *random characteristic* of the individual type. (If we were to consider "sizes" depending also upon progeny we would have to replace the last argument of χ by Ω^I.) Then $\chi(s,t,\omega)$ is the "size" at age t of an s-type individual, leading the life ω. We assume it to be zero for negative t. At time t after population start an individual $x \in I$ will have age $t - \sigma_x$, be of type ρ_x, and lead the life U_x. Thus its "size" is $\chi(\rho_x, t - \sigma_x, U_x)$. (In the general, non-individual case U_x would be replaced by S_x.) Population "size" at t is then

$$z_t^\chi = \sum_{x \in I} \chi(\rho_x, t - \sigma_x, U_x).$$

It is to this expression the compositions mentioned are applied to yield formulas like

$$z_t^\chi = \sum_{x \text{ precedes } \mathcal{I}} \chi(\rho_x, t - \sigma_x, U_x) + \sum_{x \in \mathcal{I}} z_{t-\sigma_x}^\chi \circ S_x.$$

The choice $N = \mathcal{I}$ in the equation has the advantage that only one x precedes the line and the first sum reduces to just $\chi(\rho_0, t, U_0)$. The asymptotics of $e^{-\alpha t} E_s[z_t^\chi]$ then follows under assumptions that make Markov renewal theory work. We refrain from spelling those out, referring for general such theory to Niemi and Nummelin, 1986, and for precise results on branching populations to Jagers, 1988.

"Theorem" 4 *For decent reproduction kernels and characteristics*

$$\lim_{t\to\infty} e^{-\alpha t} E_s[z_t^\chi] = h(s) E_\pi[\hat{\chi}(\alpha)]/\alpha\beta.$$

Here

$$\beta = \int_{S\times R_+} th(s) e^{-\alpha t} \mu(\pi, ds \times dt),$$

assumed finite and positive, is (or can be shown to be) the *backwards genera-tion span* or in the one type case the *mean age at child-bearing*, and $\mu(\pi, \cdot) = \int \mu(s, \cdot)\pi(ds)$. "Decent" actually means that the kernel should be what is called spread out and that the characteristic may not be to unbounded.

Only few assumptions - but hard work! - must be added for the step from expected to actual population behaviour. What is needed is the $x \log x$-assumption, yielding the L^1-limit w of the intrinsic martingale, $\inf h > 0$, that no individual could have infinitely many children, and that the total population size, $y_t = \#\mathcal{Y}_t$, at fixed t should not be too terribly large for some starting types - technically that for each t y_t should be uniformly integrable with respect to $P_s, s \in S$.

"Theorem" 5 *Under the above assumptions*

$$e^{-\alpha t} z_t^\chi \to h(s) w E_\pi[\hat{\chi}(\alpha)]/\alpha\beta,$$

in the mean for π-almost all starting types $s \in S$, as $t \to \infty$. Moreover $w = 0$ precisely when the population becomes extinct.

Since, for $w \neq 0$, two such limits can be divided, this solves the asymptotic composition problem for non-extinct populations.

References

1. B. Charlesworth, 1980, *Evolution in Age-structured Populations.* Cambridge University Press.

2. B. Chauvin, 1986, Arbres et processus de Bellman-Harris. *Ann. Inst. H. Poincaré* 22, 209 -232.

3. P. Jagers, 1975, *Branching Processes with Biological Applications.* John Wiley and Sons.

4. P. Jagers, 1988, General Branching Processes as Markov Fields. *Submitted.*

5. P. Jagers and O. Nerman, 1984, The growth and composition of branching populations. *Adv. Appl. Prob.* 16, 221 - 259.

6. P. Jagers and O. Nerman, 1985, Branching processes in periodically varying environment. *Ann. Prob.* 13, 254 - 268.

7. O. Kallenberg, 1983, *Random Measures*, 3rd ed. Akademie-Verlag/Academic Press.

8. N. Keyfitz, 1985, *Applied Mathematical Demography*, 2nd ed. Springer-Verlag.

9. J. A. J. Metz and O. Diekmann (eds.), 1986, *The Dynamics of Physiologically Structured Populations.* Lecture Notes in Biomathematics 68, Springer-Verlag.

10. O. Nerman, 1981, On the convergence of supercritical general (C-M-J) branching processes. *Z. Wahrscheinlichkeitstheorie verw. Geb.* 57, 365 - 395.

11. O. Nerman, 1984, *The Growth and Composition of Supercritical Branching Populations on General Type Spaces.* Dep. Mathmematics, Chalmers U. Tech. and Gothenburg U., 1984-18.

12. S. Niemi and E. Nummelin, 1986, On non-singular renewal kernels with an application to a semigroup of transition kernels. *Stoch. Proc. Appl.* 22, 177 - 202.

Acta Applicandae Mathematicae **14** (1989), 115–123.

On A Generalized Mathematical Model of the Immune Response

Miloš Jílek

Institute of Microbiology

Czechoslovak Academy of Sciences

14220 Prague 4

Czechoslovakia

AMS Subject Classification (1980): 92A07
Key words: immune system, cell differentiation, modelling

Immune response is one of the most important and most interesting biological processes studied by contemporary biology.

Mathematical models of this process are based on different assumptions. As mentioned by Hoffmann et al. (1986), first steps of the construction of mathematical models are relatively simple and easy when biological postulates are formulated properly, but many difficult mathematical problems (such as the existence, uniqueness and stability of solutions, etc...) are to be solved in connection with their further developing. These problems are being successfully solved in connection with several models (see, e.g., Marchuk et al. 1986, Belykh 1983, Zuev 1986, Klaschka 1988).

However, there exist classes of models based on (more or less) similar assumptions, and formulation of more general models could facilitate the most effective solution of some problems connected with particular models.

An attempt at the formulation of a generalized model of the course of the immune response is given in this paper.

First, mathematical models of the immune response were based on the scheme suggested by Sercarz and Coons (1962) at a conference held in Prague 27 years ago (see Fig. 1). This scheme, including a two-stage differentiation process of cells taking part in the immune response, was modified by Šterzl (1967) - see Fig. 2 - and this new version led to the construction of first mathematical models of the immune response (Jílek and Šterzl 1970, Bell 1970). This X-Y-Z scheme has been very fruitful - many contemporary models of the immune response are more or less based on it (see Jílek and Přikrylová 1985, 1987).

Therefore, it seems possible to make a simple generalization of models based on the X-Y-Z scheme (it means that models based on the network theory will not be involved in this generalized model).

Most of the contemporary models are constructed as systems of differential equations, and such is even the generalized model (it means that stochastic processes constructed by ourselves in the seventies (Jílek and Šterzl 1970, 1973, Jílek 1973, 1975, Jílek and Klein 1978) as well as systems of logical equations given by Přikrylová and Kurka (1984) and Kaufman et al. (1985) are not included in this generalized model).

Several models mentioned in the present paper consider B cells, T cells and macrophages as well as their cooperation; for simplicity, an attempt to generalize equations for the amount of antigen, antibody and their complexes only will be done; a generalized form of equations for other components of the immune system will be given in a more detailed study which is in preparation.

In the following survey of equations, some multiplying coefficients (as well as the time delay considered in several models) are omitted, and several equations are simplified. (Of course, the survey is not complete, and many other papers using similar equations describing the course of appropriate components of the immune system could be quoted.)

The following notation is used in the generalized model:

Signals and products

antigen	Ag (Ag_f, Ag_b - free, bound antigen)
antibody	Ab (Ab_f, Ab_b - free, bound antibody)
interleukin 1	$IL1$
growth factor	GF

Cells

macrophages	M	
lymphocytes	B (B cells)	H (T helper cells)
immunocompetent cells	X	H_x
proliferating cells	Y (Y_1, Y_2)	H_y
memory cells	Y_m	H_m
producing cells	Z (Z_1, Z_m)	H_z

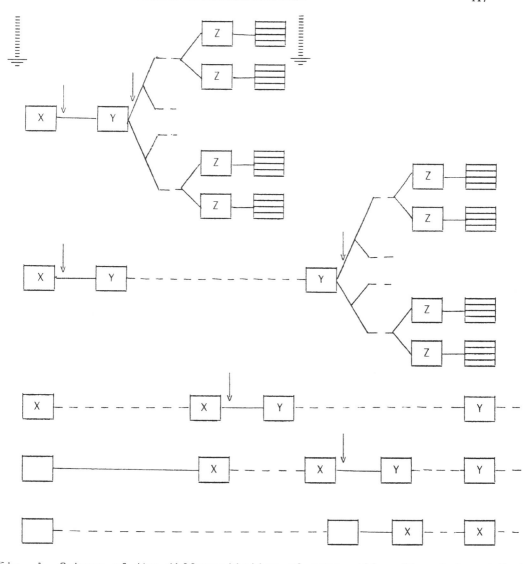

Fig. 1. Scheme of the differentiation of some cells after 1st and 2nd injection of antigen; open squares - precursors of X cells, hatched squares - eliminated cells, thick arrows - injection of antigen, thin arrows - cellular contacts with antigen (after Sercarz and Coons 1962, simplified).

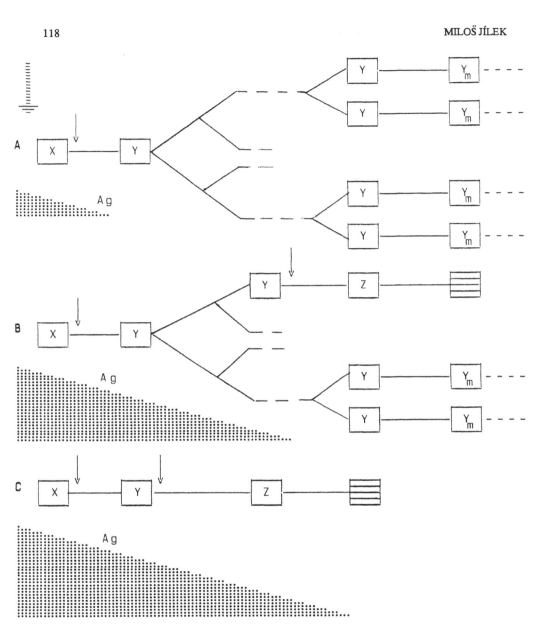

Fig. 2. Scheme of the differentiation of some cells after injection
antigen; **A** - priming (memory formation), **B** - antibody production and
memory formation, **C** - exhaustive terminal differentiation, Ag - amou
of antigen; for other symbols see Fig. 1. (after Šterzl 1967, simpli
fied).

Complexes C $(C_{Ag,Ab}, C_{Ag,\text{cells}})$

Rates (with appropriate indexes)

birth l
death m
bounding b
decay d
source s

Functions (not discussed here) F

Constants a_x, a_y, k, N, p, q

Let us compare some equations concerning the time course of the amount of free antigen after its inoculation into the organism:

$Ag_f'(t) = m_{Ag}'(t)Ag(t)$, where $m_{Ag}(t)$ is a real function

$\quad$ (e.g., $m_{Ag}(t) = m_{Ag,i}$, $t \in [t_{i-1}, t_i)$, $Ag_f(t_0) = a_0$,

$\quad$ where $m_{Ag,i}$, t_i are given nonnegative real numbers,

$\quad$ $i = 1, \ldots, k$ (k given), $0 = t_1 < t_2 < \cdots < t_k = \infty$

$\quad$ - example:

$$Ag_f(t) = a_0 \exp\left(-\sum_{j=1}^{i-1}(m_{Ag,j} - m_{Ag,j+1})\exp(-m_{Ag,i}t)\right),$$

$\quad$ $t \in [t_{i-1}, t_i)$;

$\quad$ Jílek and Šterzl (1970, 1973) and Klein et al. (1981) used $k = 1$,

$\quad$ Přikrylová et al. (1984) used $k = 3$

$Ag_f'(t) = (l_{Ag} - p \cdot b_{Ag,Ab}Ab(t))Ag_f(t)$

$\qquad\qquad$ **(Marchuk 1983 - simple model, Kaufman et al. 1985, Přikrylová 1986, 1988)**

$Ag_f'(t) = s_{Ag} - m_{Ag}Ag_f(t) + k \cdot d_{C_{Ag,Ab}}Ag_b(t)$

$\qquad$ (Bell 1970)

$$Ag_f'(t) = -b_{Ag,Ab}Ag_f(t)Ab_f(t) + d_{C_{Ag,Ab}}C_{Ag,Ab} - (a_xX(t) + a_yY(t))Ag_f(t)$$

(Grossman et al. 1980)

$$Ag_f'(t) = s_{Ag}(t) - m_{Ag}Ag_f(t) - N(b_{Ag,Ab}Ag_f(t)Ab_f(t) - d_{C_{Ag,Ab}}C_{Ag,Ab}(t))$$

(Mohler et al. 1980)

These (and some other) equations for the time course of the amount of free antigen indicate the possibility to construct more general equations, e.g.,

$$Ag_f'(t) = s_{Ag}(t) - m_{Ag}(t)Ag_f(t) - b_{Ag,Ab}Ag_f(t)Ab_f(t) + d_{C_{Ag,Ab}}C_{Ag,Ab}(t)$$
$$- \sum_{\text{cells}} b_{Ag,\text{cells}}Ag_f(t).\text{cells}\,(t) + \sum_{\text{cells}} d_{C_{Ag,Ab}}C_{Ag,Ab}(t)$$

Similarly, we can compare some equations describing the time course of the amount of free antibodies:

$$Ab_f'(t) = l_{Ab}Z(t) - m_{Ab}Ab(t)$$
(Marchuk 1983 - simple model)

$$Ab_f'(t) = l_{Ab}Z(t) - b_{Ag,Ab}Ag(t) - m_{Ab}Ab(t)$$
(Marchuk and Petrov 1983, Příkrylová 1986, Příkrylova et al. 1986)

$$Ab_f'(t) = l_{Y:Ab}Y(t) + l_{Z:Ab}Z(t) - b_{Ag,Ab}Ag(t)Ab(t) - m_{Ab}Ab(t)$$
(Bell 1970)

$$Ab_f'(t) = l_{X:Ab}X(t) + l_{Z:Ab}Z(t) - m_{Ab}Ab(t) - b_{Ag,Ab}Ag_f(t)Ab_f(t)$$
$$+ d_{C_{Ag,Ab}}C_{Ag,Ab}(t)$$

(Mohler et al. 1980)

$$Ab_f'(t) = k.F_1(Ab(t),Z(t))F_2(Ab(t), Ag(t))F_3(Ab(t), H(t))$$

$$- q . b_{Ag,Ab} Ag(t) Ab(t) - m_{Ab} Ab(t)$$

(Kaufman et al. 1985)

A comparison of these equations leads to the following more general equation:

$$Ab_f'(t) + \sum_{\text{cells}} l_{\text{cells}:Ab} \text{cells}(t) + d_{C_{Ag,Ab}} C_{Ag,Ab}(t) - b_{Ag,Ab} Ag_f(t) Ab_f(t) - m_{Ab} Ab(t)$$

For complexes $C_{Ag,Ab}$, the following equation should be considered:

$$C_{Ag,Ab}'(t) = b_{Ag,Ab} Ag_f(t) Ab_f(t) - (d_{C_{Ag.Ab}} + m_{C_{Ag.Ab}}) C_{Ag,Ab}(t)$$

(Mohler et al. 1980, Grossman et al. 1980)

Similar reasons lead to general equations also for further components of the immune system, which will be given in a more detailed study which is in preparation.

Future work concerning the modeling of the immune response should concentrate on the study of functions which can be considered as switching (or transition) functions supplying the transition probabilities; it would also be useful to critically assess the use of complicated functions which lead to the nonlinearity (or nonbilinearity) of differential equations.

References

Bell, G.I., Mathematical model of clonal selection and antibody production. *Nature 228* (1970), 739-744; *J. Theoret. Biol. 29* (1970), 191-232; *33* (1971), 339-378; 379-398.

Belykh, L.N., On the computation methods in disease models. pp. 79-84 in: *Mathematical Modeling in Immunology and Medicine.* North-Holland, Amsterdam - New York - Oxford 1983.

Grossman Z., Azofsky, R., DeLisi, C., The dynamics of antibody secreting cell production: Regulation of growth and oscillations in the response to T- independent antigens. *J. Theoret. Biol. 84* (1980), 49-92.

Hoffmann, G.W., Cooper-Willis, A., Chow, M., On paradoxes and progress in theoretical immunology, and evidence for a new symmetry. pp. 15-31 in: *Immunology and Epidemiology*. Springer-Verlag, Berlin - Heidelberg - New York - Tokyo 1986.

Jílek, M., Immune response and its stochastic theory. pp. 209-212 in: *Identification and System Parameter Estimation*. North-Holland, Amsterdam 1973.

Jílek, M., Stochastic development of cell populations under non-homogeneous conditions. *Acta Biotheoretica 24* (1975), 108-119.

Jílek, M., Klein, P., Stochastic model of the immune response. pp. 15-25 in: *Modelling and Optimization of the Complex System*. Springer-Verlag, Berlin - Heidelberg - New York 1978.

Jílek, M., Přikrylová, D., The X-Y-Z scheme after 23 years. *Folia Microbiol. 30* (1985), 302-311.

Jílek, M., Přikrylová, D., The X-Y-Z scheme as a basis for modelling and simulation of the immune response. pp. 197-201 in: *European Congress on Simulation, Vol. A* Academia, Prague 1987.

Jílek, M., Šterzl, J., A model of differentiation of immunologically competent cells. pp. 963-981 in: *Developmental Aspects of Antibody Formation and Structure*. Academia, Prague 1970.

Jílek, M., Šterzl, J., On a theory of the immune response. pp. 275-289 in: *Trans. 6th Prague Conf. Information Theory, Statist. Decision Functions, Random Processes*. Academia, Prague 1973.

Kaufman, N., Urbain, J., Thomas, R., Towards a logical analysis of the immune response. *J. Theoret. Biol. 114* (1985), 527-561.

Klaschka, J., Hill functions of n variables in models of cell kinetics (in preparation).

Klein, P., Šterzl, J., Doležal, J., A mathematical model of B lymphocyte differentiation. *J. Math. Biol. 13* (1981), 67-86.

Marchuk, G.I., *Mathematical Models in Immunology*. Optimization Software, Inc., Publications Div., New York 1983.

Marchuk, G.I., Asachenkov, A.L., Belykh, L.N., Zuev, S.M., Mathematical modelling of infectious diseases. pp. 64-81 in: *Immunology and Epidemiology*. Springer-Verlag, Berlin - Heidelberg - New York - Tokyo 1986.

Marchuk, G.I., Petrov, R.V., The mathematical model of the anti-viral immune response. pp. 161-173 in: *Mathematical Modeling in Immunology and Medicine*. North-Holland, Amsterdam - New York - Oxford 1983.

Mohler, R.R., Bruni, C., Gandolfi, A., A system approach to immunology. *Proc. IEEE 68* (1980), 964-990.

Přikrylová, D., Mathematical modelling of the immune response: A model of the proliferation control. pp. 44-52 in: *Immunology and Epidemiology.* Springer-Verlag, Berlin - Heidelberg - New York - Tokyo 1986.

Přikrylová, D., IL 2 and immune response control. Mathematical model (in this volume).

Přikrylová, D., Jílek, M., Doležal, J., A model of proliferation control in immune response. *Kybernetika 20* (1984), 37-46.

Přikrylová, D., Jílek, M., Doležal, J., A new mathematical model of proliferation control during immune response. *Immunol. Letters 13* (1986), 317-321.

Přikrylová, D., Kurka, P., Modelling of the immune response by means of a nondeterministic dynamic system. pp. 725/1-5 in: *Simulation of Systems in Biology and Medicine '84.* DT ČSVTS, Prague 1984.

Sercarz E., Coons A.H., The exhaustion of specific antibody producing capacity during a secondary response. pp. 78-83 in: *Mechanisms of Immunological Tolerance.* Academia, Prague 1962.

Šterzl, J., Factors determining the differentiation pathways of immunocompetent cells. *Cold Spring Harbor Symp. Quant. Biol. 32* (1967), 493-506.

Zuev, S.M., Estimation of parameters of models of the immune response (in Russian). pp. 298-308 in: *Mathematical Modeling in Immunology and Medicine* (Russian Edition). Mir, Moskva 1986.

Acta Applicandae Mathematicae **14** (1989), 125–133.

Recent Results in Mathematical Modeling of Infectious Diseases

L.N. Belykh

Department of Numerical Mathematics

USSR Academy of Sciences

11 Gorky st., 103009 Moscow, USSR

AMS Subject Classification (1980): 92A07
Key words: stationary solution, immune response, stability

Some results connected with a simple mathematical model of infectious disease are discussed in order to demonstrate the approach to the modelling of such real processes. A more complicated model of antiviral immune response is presented. A new modification of this model in which targets for the viruses are immunocompetent cells is suggested.

1. A Simple Model and Some Results

Here we would like to demonstrate our approach to the modeling of infectious diseases and use for this aim the simple mathematical model advanced by G.I. Marchuk [2]. And at first let us define the term 'infectious disease'. What does it mean? How do we understand it?

According to medical literature the infectious disease is regarded as expression of the relationship between two members of a biocenosis, one of which (stimulant) being capable of existing in the other owing to pathogenic mechanisms, and this other organism being capable of counteracting this pathogenic action. It is also well known that the immune system plays an important role in the defense of organisms from the infections. Based on these two points we consider the infectious disease as a conflict between multiplying pathogenic antigen and the immune system, and as a first step we distinguish the following main characteristics of a disease.

1 Concentration of viruses $V(t)$. By viruses we mean multiplying pathogenic antigen.

2 Concentrations of antibodies $F(t)$. By antibodies we mean the substrates of the immune system, neutralizing viruses (immunoglobulins, cell receptor, killers).

3   Concentration of plasma cells $C(t)$. This is the population of carriers and producers of antibodies (immunocompetent cells and immunoglobulin producers).

4   Relative characteristic (mass or area) of a damaged organ $m(t)$.

Based on our view of infectious diseases and using these four variables a simple mathematical model a of disease was constructed in the form of the system

$$\frac{dV}{dt} = (\beta - \gamma F) V$$

$$\frac{dF}{dt} = \rho C - \eta \gamma F V - \mu_f F$$

$$\frac{dC}{dt} = \xi(m) \cdot \alpha V(t-\tau) F(t-\tau) \theta(t-\tau) - \mu_c (C - C^*) \tag{1}$$

$$\frac{dm}{dt} = \sigma V - \mu_m m$$

with initial condition at $t = t^o = 0$

$$V(0) = V^o \geq 0, \ \ F(0) = F^o \geq 0, \ \ C(0) = C^o \geq 0, \ \ m(0) = m^o \geq 0 \ . \tag{2}$$

(Here $\theta(t)$ is the Heavyside function: $\theta(t) = 0$ at $t < 0$, $\theta(t) = 1$ at $t \geq 0$).

Details of the model construction can be found in [1,2]. Here we only describe the disease process in accordance to this model. At some moment of time $t = t^o = 0$ a small population of viruses V^o penetrates the healthy organism. This population begins to multiply in the cells of an organ and thus to damage the organ. Part of the viruses binds the receptors immunocompetent cells and such contact stimulates these cells to divide and proliferate. After some time interval τ necessary for cell division and proliferation, the clone of plasma cells arises. Plasma cells produce with high speed the antibodies which neutralize the viruses. The outcome of the disease depends on the struggle between them. If the viruses damage the target organ seriously, the efficiency of the immune system response is getting worse and the possibility of the recovery becomes less. (It is described by the function $\xi(m)$ which is supposed to be a non-negative non-increasing function of limited argument $m (0 \leq m \leq 1)$ with the properties $\xi(0) = 1$, $0 \leq \xi(m) \leq 1$, $\xi(1) = 0$.)

The model does not describe a specific disease caused by a specific antigen. Our main objective of modeling was to describe and find general laws inherent to all infectious diseases. With this aim we first of all established the three main features of global adequacy of our model to real processes which are formulated below in the form of theorems.

Theorem 1. For all $t \geq t^o = 0$ there exists a unique solution of system (1) with initial condition (2).

Theorem 2. This solution is non-negative for all $t \geq t^o = 0$.

Theorem 3. The inequality $\beta < \gamma F^*$ is a sufficient condition for the asymptotical stability of the stationary solution

$$V_1 = 0, \quad F_1 = F^* = \rho C^* / \mu_f, \quad C_1 = C^*, \quad m_1 = 0, \tag{3}$$

and, moreover, if $F^o = F^*$, $C^o = C^*$, $m^o = 0$ and

$$0 < V^o < V^* = \frac{\mu_f(\gamma F^* - \beta)}{\beta \eta \gamma} \tag{4}$$

then $V(t)$ decreases on the interval $[0, \infty)$ and

$$V(t) \leq V^o e^{-at}$$

where

$$a = \frac{\gamma \rho C^*}{\mu_f + \eta \gamma V^o} - \beta > 0.$$

We described these theorems as the main features of global adequacy because they are essential to any model of a real infectious process. Indeed, for example, Theorem 1 guarantees that the biologically absurd event is impossible, i.e., that the quantity of any process component does not become infinite in finite time (global existence). Secondly, it guarantees repeatability of experiments, i.e. the same conditions of biological experimentation must lead to the same results (global uniqueness). Theorem 2 guarantees nonnegativity of the model components as the biology of the process dictates. Finally Theorem 3 establishes the possibility for the healthy body state (i.e. the stationary solution (3)) to be stable and in this case guarantees the existence of an immunologic barrier V^* (see (4)) such that in a healthy body infected by a small viral dose $V^o < V^*$ the disease does not develop. In other words if the immunologic barrier of the healthy body is sufficiently high then it is not necessary for this organism to be ill after infection by a small viral dose. This is a very natural situation because in our life we have a lot of contacts with different viruses (penetrating the organism with air, food and so on) but we are not always ill. Moreover in the majority of cases we continue to be healthy.

As far as Theorem 3 is concerned we point out here one biological corollary. Namely, the raise in the level of the immunocompetent cells in a healthy organism, C^*, raises the immunologic barrier and therefore is an effective method of preventing and, possibly, treating the disease. This corollary is confirmed by the well known vaccination procedure which due to the formation of memory cells raises substantially the level of the immunocompetent cells in the body.

It should be stressed that we obtained from our mathematical analysis of the model not only this clear biological conclusion but a number of them. In particular we proved the following Theorem 4.

Theorem 4. A sufficient condition for the stationary solution

$$V_2 = \frac{\mu_c(\mu_f\beta - \gamma\rho C^*)}{\beta(\alpha\rho - \mu_c\eta\gamma)}, \quad C_2 = \frac{\alpha\mu_f\beta - \eta\mu_c\gamma^2 C^*}{\gamma(\alpha\rho - \mu_c\eta\gamma)}, \tag{5}$$

$$F_2 = \beta/\gamma, \quad m_2 = \frac{\sigma}{\mu_m}V_2$$

to be asymptotically stable at $\xi(m)\equiv1$ and $\alpha\to\infty$ is that the inequality

$$0 < \beta - \gamma F^* < \frac{1}{\tau + (\mu_c + \mu_f)^{-1}} \tag{6}$$

is satisfied for $\mu_c\tau\leq1$.

The condition $\mu_c\tau\leq1$ is immunologically reasonable since the time of formation of the plasma cell clone $\tau \approx 1$ day and the lifetime of a plasma cell $\tau_c = 1/\mu_c \approx 2$ to 4 days. The condition $\alpha\to\infty$ means a very high sensitivity of the immune system to the given viruses (see model (1)). So, from the mathematical Theorem 4 we can conclude that an organism having an arbitrarily high sensitivity of the immune system can contain a negligible stable non-zero viral concentration. In fact, for an arbitrarily large α we can "select" a virus with multiplication rate β such that inequality (6) is satisfied. The smallness of the viral concentration follows from the fact that $V_2\to0$ as $\alpha\to\infty$.

Our next step was numerical experimentation with model (1), or simulation. We simulated the infection of a healthy body by a small viral dose, i.e. the initial conditions were

$$V(0) = V^o > 0, \quad F(0) = F^*, \quad C(0) = C^*, \quad m(0) = 0.$$

The simulation enabled us to distinguish four qualitatively different types of model solutions which were interpreted as forms of the course of the disease. These forms are subclinical, acute with recovery, chronic and lethal. The description of these forms will be given below. Here I would like to formulate the main biological corollary. Namely, the occurrence of any form of disease by the infection of a healthy body by some small viral dose is independent of this dose but is determined by the immunologic status of the body (the set of model parameters).

Now we describe shortly and in biological terms the results of our model analysis connected with the disease forms in question. (For details in mathematical analysis and biological interpretations see [1,2].)

The subclinical form of disease is characterized by a stable elimination of viruses from the organism and resembles the vaccination by nonpathogenic multiplying antigens, which can be interpreted as using live vaccine. In this case the organ is practically intact, and the viral concentration tends to zero in time. The immune system stimulation leads to immune response and memory cells appear. The model does not describe this effect but we can regard it in such a way that the latter elevates the level C^*, which in turn raises the immunologic barrier in the organism against viruses of a given type.

The acute form of the disease is characterized by the corresponding dynamics of the viruses: rapid growth during several days and sharp decline to zero. An effective immune response occurs in the organism leading to recovery. Only considerable damage of the organ can lead the acute form to the chronic form or to a lethal outcome. Hence the treatment of acute forms needs to be directed toward the suppression of pathogenic (but not antigenic) properties of viruses. Treatment causing the reduction of the antigenic property of the viruses, that is their ability to bring out the immune response, facilitates conversion of the acute form to the chronic one, since such treatment weakens the stimulation of the immune system by viruses.

The chronic form of the disease in case of slight damage of the organ is characterized by the presence of a nonzero population of viruses possessing flaccid dynamics in the organism. They are caused by an insufficiently effective stimulation of the immune system. In some cases they should be treated by aggravating the disease. The aggravation of the disease (i.e the substantial increase in viral concentration) leads to the effective stimulation of the immune system and therefore to a strong immune response which in turn leads to the recovery due to complete neutralization of the viruses. Based on this result we advanced the theory of biostimulation and constructed its model. The sense of this theory is the following. In the body subjected to a stable chronic form the new non-multiplying nonpathogenic antigen (biostimulant) is injected during some interval of time. So, we have in the body two antigens: "chronic" (stimulant of disease) and biostimulant. Injecting large doses of biostimulants, we can achieve a situation where the immune response to the chronic antigen will be blocked due to the competition between two antigens for the nonspecific macrophages. This means that viruses have the opportunity to multiply without any control from the immune system side. At some moment of time the injections of biostimulants are stopped and they are eliminated from the body by antibodies specific for them. The organism is again faced with the chronic antigen. But the situation has essentially changed. During the interval of aggravation (when the biostimulants were injected) the virus concentration reached a level which leads to the effective stimulation of the immune system. A strong immune response arises and recovery follows.

The lethal outcome of a disease is connected with severe (complete) damage of an organ which is no longer capable of securing a normal vital activity of the organism. Severe damage of the organ is caused either by a high pathogeneity of the viruses (the coefficient σ is large) or by weak (small stimulation coefficient α) and untimely (long time of plasma cells clone formation τ) immune response. Both acute and chronic forms can lead to a lethal outcome in the case of severe damage of the organ.

So, this simple model of a disease enabled us to distinguish and characterize forms of disease from an immunological point of view. It should be noted that the model was modified for the description of such processes as the influence of temperature reaction on the infectious disease, the course of a mixed disease (biinfection), the action of the factor SAP (stimulator of antibody producers) and many others.

2. Mathematical Model of Antiviral Immune Response

The further development and generalization of the simple model was the model of antiviral immune response advanced by Academicians G.I. Marchuk (a mathematician) and R.V. Petrov (an immunologist). This model has a form

$$\dot{V}_f = nb_E C_V E + pb_m C_V - \gamma_m M V_f - \gamma_f V_f F - k_V \sigma C V_f$$

$$\dot{M}_V = \gamma_M M V_f - \alpha_M M_V$$

$$\dot{H}_E = b_H [\xi(m)\rho_H M_V H_E \big|_{t-\tau_H} - M_V H_E] - b_P M_V H_E E + \alpha_H (H_E^* - H_E)$$

$$\dot{H}_B = b_H^{(B)} [\xi(m)\rho_H^{(B)} M_V H_B \big|_{t-\tau_H^{(B)}} - M_V H_B] - b_P^{(B)} M_V H_B B + \alpha_H^{(B)} (H_B^* - H_B)$$

$$\dot{E} = b_P [\xi(m)\rho_E M_V H_E E \big|_{t-\tau_E} - M_V H_E E] - b_E C_V E + \alpha_E (E^* - E) \tag{7}$$

$$\dot{B} = b_P^{(B)} [\xi(m)\rho_B M_V H_B B \big|_{t-\tau_B} - M_V H_B B] + \alpha_B (B^* - B)$$

$$\dot{P} = b_P^{(P)} \xi(m)\rho_B M_V H_B B \big|_{t-\tau_B} + \alpha_P (P^* - P)$$

$$\dot{F} = \rho_f P - \eta_f \gamma_f V_f F - \alpha_f F$$

$$\dot{C}_V = \sigma C V_f - b_E C_V E - b_m C_V$$

$$\dot{m} = \mu b_E C_V E + \eta b_m C_V - \lambda m$$

with appropriate initial condition.

In principle this system describes the same process which we studied before but the description is more detailed from the immunological point of view. First of all the antigen exists here in two forms: free viruses (V_f) which are freely circulating in the body and viruses (C_V) in infected organ's cells. The immune system also has two subsystems: the T-system which includes helpers for humoral (H_B) and cell-mediated (H_E) responses and killers (E), and the B-system which includes B-cells (B) and plasma cells (P). The antigen-presenting cells are macrophages stimulated by free viruses (M_V). As before, F means antibodies and m is some characteristic of organ damage. So, this model describes the cooperation of three cellular systems: T-, B- systems and macrophages. For the stimulation of helpers H_B and H_E only one signal from the macrophages M_V is required while for the stimulation of B-cells B and effectors E two signals are required: from the macrophages M_V and the corresponding helpers H_B or H_E. The immune system responds to antigen invasion in two ways. The first is the humoral response through the production of antibodies F which neutralize only free viruses V_f. The second is the cell-mediated response with accumulation of effector or killer cells E which can destroy only the organ's cell infected by the viruses. The free viruses V_f originate from the death of infected cells C_V due to both irreversible processes inside the cell and the killing of infected cells by infectors E. I think this is enough in order to imagine the general picture described by the model (For details see [2].)

The corresponding theorems on global existence, uniqueness and nonnegativity as well as stability of the healthy state were proved. The numerical analysis of this model confirmed the main biological conclusions of the simple model (1) and pointed out the leading role of the T-system in the antiviral defense. The new fact discovered by this model was the possibility of a situation where the very aggressive killers can in effect lead to the death of the body. The point is that killers destroy the cells of their *own* body which is infected by viruses. On one side this leads to organ damage and on the other side this facilitates the appearance of free viruses from the cell destroyed by killers. These "new" viruses begin to infect cells again and the situation is repeated. Such "chain reactions" lead at last to the death of the body. It is possible to prevent this situation by suppression of the cell-mediated response.

This model was also used for the description of real experimental data on viral hepatitis B.

3. On a New Modification of the Marchuk-Petrov Model

Let us consider the situation when the targets for viruses V are the cells of the immune system, for example, the helpers of humoral response H_Σ. Denoting by $H_1=H_B$ the specific helpers for the given viruses and by H - the specific helpers for the other viruses we can write

$$H_\Sigma = \sum_{i=1}^N H_i = H_1 + \sum_{i=2}^N H_i = H_1 + H = H_B + H$$

where N is the number of specifities $(N \approx 10^6-10^7)$. Then in the absence of viruses we have

$$\dot{H}_\Sigma = -\mu_H(H_\Sigma-H_\Sigma^*) = \mu_H H_\Sigma^* - \mu_H H_\Sigma = k - \mu_H H_\Sigma$$

or, introducing $\epsilon=1/N$

$$\dot{H}_1 = \epsilon k - \mu_H H_1$$

$$\dot{H} = (1-\epsilon)k - \mu_H H .$$

In the presence of viruses but without immune response we have

$$\dot{H}_\Sigma = k - \mu_H H_\Sigma - \gamma H_\Sigma V$$

or

$$\dot{H}_1 = \epsilon k - \mu_H H_1 - \epsilon \gamma H_1 V$$

$$\dot{H} = (1-\epsilon)k - \mu_H H - (1-\epsilon)\gamma H V .$$

Adding the immune response according to model (7) and denoting by $H_V = C_V$ the infected helpers we obtain the model

$$\dot{H} = (1-\epsilon)k - \mu_H H - \gamma(1-\epsilon)HV$$

$$\dot{V} = nb_E H_V E + pb_m H_V - \gamma_m MV - \gamma_f FV - k_V \gamma H_\Sigma V$$

$$\dot{M}_V = \gamma_M MV - \alpha_M M_V$$

$$\dot{H}_E = b_H[\xi(m)\rho_H M_V H_E\big|_{t-\tau_H} - M_V H_E] - b_P M_V H_E E + \alpha_H(H_E^* - H_E)$$

$$\dot{H}_1 = \dot{H}_B = b_H^{(\beta)}[\xi(m)\rho_H^{(\beta)} M_V H_B\big|_{t-\tau_H^{(P)}} - M_V H_B] - b_P^{(\beta)} M_V H_B B - \epsilon\gamma H_B V + \epsilon k - \mu_H H_B$$

$$\dot{E} = b_P[\xi(m)\rho_E M_V H_E E\big|_{t-\tau_E} - M_V H_E E] - b_E H_V E + \alpha_E(E^* - E) \tag{8}$$

$$\dot{B} = b_P^{(\beta)}[\xi(m)\rho_B M_V H_B B\big|_{t-\tau_B} - M_V H_B B] + \alpha_B(B^* - B)$$

$$\dot{P} = b_P^{(\beta)}\xi(m)\rho_B M_V H_B B\big|_{t-\tau_B} + \alpha_P(P^* - P)$$

$$\dot{F} = \rho_f P - \eta_f \gamma_f F V - \alpha_f F$$

$$\dot{H}_V = \gamma H_\Sigma V - b_E H_V E - b_m H_V$$

$$\dot{m} = \mu b_E H_V E + \eta b_m H_V - \lambda m$$

$$H_\Sigma = H_B + H .$$

The simple analysis of this model displays the danger of such kind of viruses. If they have a high immunogeneity (i.e. the ability to cause a very strong immune response) then even in the case of their low pathogeneity (i.e. when infected helpers can function normally) they can cause the death of the organism from another disease, due to a highly aggressive killers effect against the helper population. In other words these viruses indirectly "open the door" for other diseases and use for this the immune system itself. From this point of view such viruses are very dangerous. In order to prevent this situation it is necessary to suppress the immune response directed against these viruses.

This model appears to be of interest because the process described by it resembles the course of AIDS.

4. References

[1] Belykh, L.N. Analysis of Some Mathematical Models in Immunology. Nauka, Moscow, 1988 (in press, Russian).

[2] Marchuk, G.I. Mathematical Models in Immunology. Optimization Software Inc., Publication Division, New York, 1983.

Acta Applicandae Mathematicae 14 (1989), 135–142.
© 1989 by *IIASA*.

Applications of the Mathematical Model of Immunological Tolerance to Immunoglobulin Suppression and AIDS

T. Hraba

Institute of Molecular Genetics
Czechoslovak Academy of Sciences
142 20 Prague, Czechoslovakia

&

J. Doležal

Institute of Information Theory and Automation
Czechoslovak Academy of Sciences
182 08 Prague, Czechoslovakia

AMS Subject Classification (1980): 92A07
Key words: immune system, modelling, simulation

1. Introduction

Experimental data on population dynamics of lymphocytes are most extensive in mice. This species has been the most widely used and the best analysed in immunological research during the last decades. Our mathematical model of immunological tolerance developed on the basis of experimental findings in tolerance induced to human serum albumin in hatched chickens [1] was applied also to tolerance in mice [2,3]. In this paper we will report applications of this model to two other categories of inhibition phenomena. One of them was experimentally induced in mice and the other observed in human infection.

(i) The first phenomenon is isotype or idiotype suppression of short duration induced in neonatal mice [4,5]. This suppression seems to be due to elimination of B cells by antibodies against the respective markers of their surface immunoglobulins.

(ii) The other inhibition phenomenon is the depletion of CD4[+] lymphocytes in persons infected with the human deficiency virus (HIV), eventually in patients suffering from the acquired immune deficiency syndrome (AIDS).

2. The mathematical model

Two developmental compartments of lymphocytes, B or T cells, are considered in the model:

(i) the immature cell compartment - $\overline{P}$ cells

(ii) the mature cell compartment - P cells.

The model is based on two assumptions: (a) antigen induces tolerance by irreversible inactivation of lymphocytes specifically reactive to it; (b) the escape from tolerance is effected by differentiation
of lymphocytes reactive to the tolerated antigen after the decrease of
the antigen in the organism. The immature antigen-reactive lymphocytes,
$\overline{P}$ cells, arise by an antigen-independent differentiation from their precursors which do not react with antigen (Fig. 1). They mature independently of antigen, too, into mature lymphocytes, P cells. The parameter
$\overline{\tau}_p$ is the rate (all rates are in day^{-1}) of maturation of $\overline{P}$ into P
cells, τ_p is the rate of natural death of P cells, and $\overline{\tau}_p\overline{P}_E$ the rate
of differentiation of $\overline{P}$ cells from their precursors (the index E denotes the steady state value). The quantity $c_p(t)$ is the rate of irreversible inactivation of P cells by the tolerizing antigen, and
analogously, $\overline{c}_pa(t)$ that of $\overline{P}$ cells. In idiotype and isotype suppression antibodies against the respective markers of immunoglobulins assu-
me the role of antigen in tolerance induction. Actually, the immunoglobulin cell receptors bind the antigen in B cell tolerance, and the injected antibodies bind to immunoglobulin B cell receptors in idiotype
and isotype suppression. It is probable that the mechanism of the resulting inhibition of B cells is in both cases the same.

Sizes of $\overline{P}$ and P cell compartments are described by the following differential equations with the given initial values:

$$d\overline{P}(t)/dt = \overline{\tau}_p[\overline{P}_E - \overline{P}(t)] - \overline{c}_pa(t)\,\overline{P}(t), \quad \overline{P}(0) = \overline{P}_0, \tag{1}$$

$$dP(t)/dt = \overline{\tau}_p\overline{P}(t) - \tau_pP(t) - c_pa(t)\,P(t), \quad P(0) = P_0, \tag{2}$$

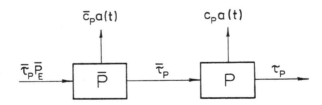

Fig. 1

where $\overline{P}(t)$ and $P(t)$ are numbers of $\overline{P}$ and P cells at time t. From the steady-state considerations in the absence of antigen, i.e. $a(t)=0$, it simply follows that $\overline{\tau}_p \overline{P}_E = \tau_p P_E$.

Denote $P_c(t)$ the number of P cells in the controls at time t, which is obtained as the solution of the model equations (1)-(2) with $a(t)=0$. Then

$$r(t) = 100 \left[P(t) \,/\, P_c(t) \right] \qquad (3)$$

is the percent measure of P cell recovery from tolerance.

3. Modelling of idiotype and isotype suppression

We simulated and compared with experimental values the recovery from the $Ac38^+$ idiotype suppression [4], and λ_1-isotype and IgM - -isotype suppression [5] induced in neonatal mice by injection of mono-clonal antibodies (mAb) Ac38, Ls136 and AF6, respectively.

The used $a(t)$ values were as follows:

$$a(t) = \begin{cases} 0, & 0 \leqslant t < t_1 \\ a_0, & t_1 \leqslant t \leqslant t_2, \\ a_0 \exp(-\beta(t-t_2)), & t \geqslant t_2, \end{cases} \qquad (4)$$

where a_0 depends on the amount of mAb injected, β is the rate of its non-imune elimination, t_1 is the day of mAb administration (the day of the birth), and t_2 is the day, when the mAb concentration starts to decreace below the suppression level a_0. In fact, time course of $a(t)$ for $t_1 \leqslant t \leqslant t_2$ does not influence the recovery from suppression, as long as $a(t) \geqslant a_0$. It serves only to simulate the retarted recovery from suppression, and for the sake of simplicity a constant value $a(t)=a_0$, $t_1 \leqslant t \leqslant t_2$ was chosen for simulation runs.

The same value of 5 day lifespan for both immature $(\overline{\tau}_p=0.2)$ and mature $(\tau_p=0.2)$ B cells was used. This value seems to be the maximal acceptable one on the basis of available experimental evidence [6]. The following parameters were used: $\overline{P}_0 = \overline{P}_E = 100$, $P_0 = P_E = 0$, $\overline{c}_p = 5$, $c_p = 1$, $a_0 = 2.4$, $t_1 = 1$. The value β designates in this case the elimination rate of the suppression inducing mAb. The elimination rate of mAb Ac38 was established to be $\beta = 0.2$ at the age of escape from suppression. We are not aware of data on elimination rates of mAb Ls136 and AF6, but as they are of the same isotype (IgG_1) as mAb Ac38, we accepted that they were the same. Therefore $\beta = 0.2$ was used in all instances in the initial simulation runs. As it was impossible to obtain satisfactory fit of these simulated curves with experimental data [7], other values were tested. The best fit in idiotype $Ac38^+$ suppression was obtained with a slower elimination rate $\beta = 0.1$ and $t_2 = 47$ (Fig. 2, curve c; crosses designa-

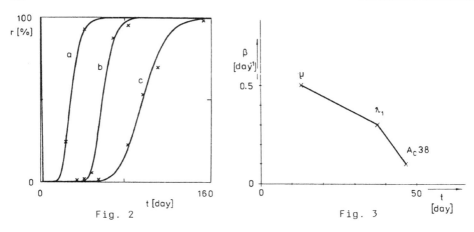

Fig. 2   Fig. 3

te experimental values). On the other hand, in λ_1 - isotype
and IgM - isotype suppression the best fit was obtained
with faster elimination rates: in λ_1 - suppression with β = 0.3 and
t_2 = 38 (Fig. 2, curve b), and in IgM - suppression with β = 0.5
and t_2 = 13 (Fig. 2, curve a).

There is a substantial difference in age when these recoveries
from suppression occur. In Ac38[+] - suppression, the recovery starts at
the age of 7 weeks, i.e. in young adult mice. The recovery from the λ_1-
- suppression starts at the beginning of the second month of age, and
from the IgM - suppression already during the first month after birth.
Figure 3 shows the age dependence of the best fitting values. It can be
concluded that the best fitting elimination rates are not related to the
actual elimination rates of antibodies effecting the suppression. The
explanation of this finding could be the loss of suppressive activity
of the administered mAb, when its concentration drops below a certain
level. Nevertheless, the kinetics of the recovery behaves as being me-
diated by the "virtual" age-dependent rate of elimination β . At this
moment, we do not know which mechanism causes this behaviour.

It would be of interest to establish, if similar relations exist
also in tolerance. There are some observations that tolerance disappears
faster in younger than in older animals [8], but the mechanisms of tolerance were not analysed in those cases sufficiently.

4. Modelling of CD4[+] lymphocyte depletion in HIV infection

The HIV envelope glycoprotein is tropic for the CD4 antigen,the
virus uses this antigen to enter the cell and the CD4[+] lymphocytes are
the preferential target cell of HIV in which the virus replicates after

entering the organism. Although virtually all aspects of immune functions were observed to be disrupted in AIDS patients, progressive depletion of CD4$^+$ lymphocytes seems to be the major defect. In contrast to
other human retroviruses, the HIV is strongly cytopathogenic. However
the cytopathogenicity of HIV can not account for the observed CD4$^+$
lymphocyte depletion, because only a small fraction of cells contains
virus in infected patients. The mechanism causing the depletion of CD4$^+$
lymphocytes seems to be either direct inhibition by soluble HIV proteins or an immune attack on these lymphocytes triggered by the HIV
[9,10].

Therefore, the dynamics of CD4$^+$ lymphocyte population in HIV infected persons can be described by equation (1) - (2), where P designates mature CD4$^+$ lymphocytes, and $\overline{P}$ the immature ones. In this case,
a(t) designates the amount of HIV products effecting the CD4$^+$ lymphocyte depletion:

$$a(t) = a_0 \exp \vartheta t, \tag{5}$$

where a_0 is the function of the infectious dose of HIV and ϑ characterizes the replication rate of the virus.

In tolerance to non-replicating antigens and in immunoglobulin
suppression, concentration of the causative agent - tolerogenic antigen or suppressive antibody, respectivelly - decreases with time and
this leads to recovery from suppression. On the other hand, the concentration of HIV products increases, as the infection progresses. In consequence of that, the CD4$^+$ lymphocyte depletion increases and its dynamics is reciprocal to that of tolerance and immunoglobulin suppression.

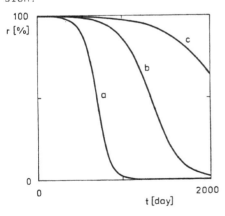

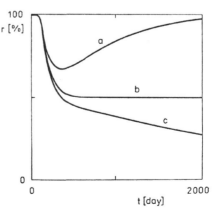

Fig. 4   Fig. 5

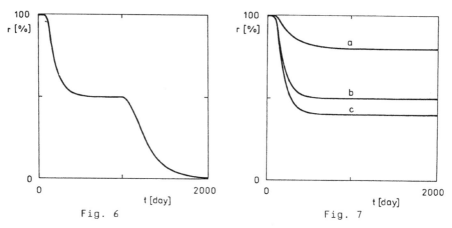

Fig. 6   Fig. 7

    In the initial runs we assumed that the immature CD4$^+$ lymphocytes
($\overline{P}$ cells) were not affected by HIV products ($\overline{c}_p = 0$). The other parameters used in these and the following runs were: $\overline{P}_o = \overline{P}_E = 2.5, P_o = P_E = 100$,
$c_p = 1.0, a_o = 10^{-5}, \overline{\tau}_p = 0.2, \tau_p = 0.005$. Simulation runs for $\vartheta = 0.1, 0.05, 0.03$
are given in Figure 4 as curves a, b, and c, respectively. Under these
conditions, significant CD4$^+$ lymphocyte depletion develops after a relatively prolonged period, but then progresses fast. However, Fahey et
al. in [11] observed substantial decrease of CD4$^+$ lymphocytes in blood
of healthy, seropositive individuals. The observed decrease remained
stationary in the majority of cases during the observation period of
18 months. A substantial drop in CD4$^+$ lymphocytes was observed in most
seroconverted individuals within several months after the appearance of
the positive serological reaction to the HIV. Evidently the CD4$^+$ lymphocyte depletion appears already in the initial phase of the HIV infection,
but then their level becomes stationary. It is possible that the arrest
of further depletion is due to immune reaction to the HIV, which arrests
further propagation of the infection, although it is unable to liquida-
te it.

    We attempted to simulate this dynamics of the CD4$^+$ lymphocyte de-
pletion in persons with HIV infection by postulating a change in
the ϑ value to ϑ' after a certain level, a_{max}, of HIV products is
reached. In Figure 5, there are given r(t) values for $\vartheta = 0.05$,
$a_{max} = 0.005$ and for different ϑ' values. For $\vartheta' = 0$, the CD4$^+$ lymphocyte depletion becomes stationary after the a_{max} value is reached
(curve b). When $\vartheta' = 5.10^{-4}$ is used, CD4$^+$ lymphocytes decrease further
but at a slower rate than in the initial phase of the infection (curve
c). The $\vartheta' = -10^{-3}$ value leads to a slow regression of the infection
with an increase in CD4$^+$ lymphocytes (curve a). For the stationary

case clearly $\bar{\tau}_p \bar{P}_E = P_E(\tau_p + c_p a_{max})$.

We are not aware of evidence, whether further progression of the HIV infection into AIDS is parallelled by a slow or abrupt decrease of $CD4^+$ lymphocytes. The latter eventuality could be effected by exacerbation of the infection due to a failure of the immune reaction which had held the progression of infection temporarily. In such a situation, the growth rate of the HIV (now equal ϑ'') would increase again. The dynamics of $r(t)$ in such a situation are given in Fig. 6. The following parameter values were used: $\vartheta = 0.05$, $a_{max} = 0.005$, $\vartheta' = 0$, $\vartheta' = 0.05$ starting to operate at day 1000 after the infection.

We assumed that immature $CD4^+$ lymphocytes are not affected by the HIV products. As CD4 antigen is present on lymphocytes early in their development, this assumption is doubtful. We have even to admit that the immature $CD4^+$ lymphocytes could be more susceptible to depletion by HIV products than the mature ones. Illustrative simulation runs taking into the account different $\bar{c}_p$ and c_p values are presented in Fig. 7. The curve b reproduces the curve b from Fig. 5 with the value $\bar{c}_p = 0$. When $\bar{c}_p = 10.0$ was used, the $r(t)$ curve did not change its character but the steady-state was reached at slightly lower values (curve c). A shift of the steady-state value to a substantially higher level was obtained with $\bar{c}_p = 10.0$ and $c_p = 0$. But again, the curve retained the same character (curve a). Thus the inclusion of immature $CD4^+$ lymphocytes in the simulation does not affect the general shape of the $r(t)$ curves, although it changes the intensity of $CD4^+$ lymphocyte depletion.

5. Conclusions

The application of our model of immunological tolerance to immunoglobulin suppression and $CD4^+$ lymphocyte depletion in HIV infected persons demonstrate that its use is not limited only to immunological tolerance and that it is also applicable to other suppression phenomena.

The results obtained in simulating the recovery from idiotype and isotype suppression suggest approaches amenable to experimental analysis and stimulating construction of alternative modifications of the mathematical model. As far as its application to the $CD4^+$ lymphocyte depletion in HIV infected individuals is concerned, we hope to be able to make it a useful tool in analysing the mechanisms of immunological defects in this infection by specifying and quantifying more precisely the involved factors.

5. References

[1]   Klein,P., Hraba,T., and Doležal J.: The use of immunological tole-
      rance to investigate B lymphocyte replacement kinetics in chickens.
      J. Math. Biology 16, (1983), 131-140.

[2]   Klein, P., Hraba, T. and Doležal J.: Mathematical model of B lym-
      phocyte replacement kinetics: its application to the recovery from
      tolerance in adult mice. Math. Biosci. 73, (1985), 227-238.

[3]   Hraba, T. and Doležal,J.: On mathematical model of immunological
      tolerance. In: System Modelling and Optimization, A. Prékopa et
      at., Eds., Springer-Verlag, Berlin 1986, 340-349.

[4]   Takemori, T. and Rajewski, K.: Specificity, duration and mechanism
      of idiotype suppression induced by neonatal injection of monoclo-
      nal anti-idiotype antibodies into mice. Eur. J. Immunol. 14(1984),
      656-667.

[5]   Saito, T., Tokuhisa, T. and Rajewsky K.: Induction of chronic idio-
      type suppression by ligands binding to the variable not the con-
      stant region of the idiotypic targent. Eur. J. Immunol. 16,(1986),
      1419-1425.

[6]   Freitas A.A., Rocha,B. and Coutinho,A.A.: Lymphocyte population ki-
      netics in the mouse. Immunol. Rev. 91, (1986), 5-37.

[7]   Doležal, J., Hraba,T.: On the choice of parameters in mathematical
      models of immunological tolerance. Proc. of the 13th IFIP Confe-
      rence on System Modelling and Optimization, Tokyo, Aug. 31 - Sept.
      4, 1987. M. Iri and K. Yajima, Eds., Springer-Verlag, Berlin, 1988.
      To appear.

[8]   Hraba, T.: Mechanism and Role of Immunological Tolerance. S.Karger,
      Basel, 1988.

[9]   Klatzmann, D., Gluckman,J.C.: HIV infection: facts and hypotheses.
      Immunology Today 7, (1986), 291-296.

[10] Ziegler, J.L., Stites, D.P.: Hypothesis: AIDS is an autoimmune
      disease directed at the immune system and triggered by a lympho-
      tropic retrovirus. Clin. Immunol. Immunopathol. 41, (1986),
      305-313.

[11] Fahey, J.L., Giorgi, J., Martínez-Maza, O., Detels, R., Mitsuyasu,
      R., Taylor,J.: Immune pathogenesis of AIDS and related syndromes.
      In: Acquired Immune Deficiency Syndrome. J.C. Gluckmann and E.Vil-
      mer (1987), 107-114.

Acta Applicandae Mathematicae **14** (1989), 143–153.
© 1989 *by IIASA*.

Distribution of Recirculating Lymphocytes:
A Stochastic Model Foundation

R.R. Mohler and Z.H. Farooqi

Department of Electrical & Computer Engineering
Oregon State University
Corvallis, OR 97331, U.S.A.

AMS Subject Classification (1980): 92A07
Key words: compartments, Markov chains, bilinear time series

Introduction

Interest in mathematical immunology has been growing. This is reflected in the several monographs that have been published and the conferences that have been held lately. Examples of recent work are available in Bell, Perelson and Pimbley (1978), Merrill (1980), Mohler, Bruni and Gandolfi (1980), DeLisi (1983), Marchuk (1983), Marchuk and Belykh (1983), Marchuk, Belykh and Zuev (1985), and Mohler (1987). The mathematical study of events that are involved at the cellular level in transmission of information seems to be generally missing in the literature, the reference here being to the study of recirculation of lymphocytes. These events functionally interconnect many of the parts of the immune response by physically and biochemically conveying information and providing defense. This paper is concerned with a particular approximation for the distribution of recirculating lymphocytes. Besides the healthy state of an organism such distribution is also of significance for disease states. Any maldistribution could be a symptom of a disease, for example, in humans, Hodgkin's disease, hepatitis B, psoriasis, and chronic lymphocytic leukemia. Thus quantification of the norm of distribution pattern, statistical analysis of the deviation from the norm and/or maldistribution, and modeling and identification of the models can be useful in diagnosis and estimation of disease severity, and in the classification and differential diagnosis of abnormalities in migratory patterns of lymphocytes.

Circulatory Lymphocyte Model Development

The most important factors that influence lymphocyte migration are i) hemo-
dynamic and ii) physico-chemical interaction between cells and vascular endothelium
(Ford, Smith, and Andrew, 1978).

The whole blood behaves as a non-Newtonian fluid with the blood viscosity
varying with the hematocrit (the percent of blood volume occupied by cells). Blood
flow in large vessels (i.e., those with diameter much larger than cell diameter) may
be considered as homogeneous and obeying Hagen-Poiseuille's law; while microcircula-
tion is characterized by low Reynold's number and non-homogeneous flow with two
phases (one plasma and the other cells). In the capillaries there is a continuous
variation in blood flow velocity and the hematocrit is unsteady (Fung, 1984).

While the erythrocytes tend to flow along the tubal axis, the leukocytes
(including lymphocytes) prefer to stick to the vascular endothelium and roll along
it (DeSousa, 1981). This causes a shear on the leukocyte which can be expressed as
a nonlinear function of the diameter of white blood cells, WBC, considered as a
sphere, diameter of blood vessel, maximum velocity of undisturbed flow in the
vessel, linear velocity of centroid of the WBC, hematocrit, Reynold's number, plasma
density, and plasma coefficient of viscosity (Fung, 1984).

This shear is strongly dependent on hematocrit which is stochastic in vivo and
consequently the shear fluctuates in time.

A leukocyte adhering to the endothelium is subjected to shear due to blood flow
and also to an effective frictional drag. Thus, accelerations and cell concentrations are functions of distance along the vessel axis, thus there results a concentration gradient. Then by Fick's first law of diffusion, a flow exists which is
proportional to concentration differences across the interface. The coefficient of
multiplication depends on the shear and drag and is thus stochastic in nature.

For the purposes of constructing the stochastic model, blood, lungs, spleen, and
bone marrow are separate compartments. All the other organs are lumped into three
compartments, assuming homogeneity of structure and function within a compartment:
i) tissues like gut, liver, Peyer's patches and miscellaneous nonlymphoid tissue as
"other tissue," ii) the lymph nodes (LN) draining the other tissues as "LN-a," and
iii) all the other LN as "LN-b."

Because of the presence of inherent stochasticity and the lack of precise
knowledge of various variables, as noted above, the model in discrete-time vector
form becomes from Fick's law and cell conservation:

$$x(n+1) = Ax(n) + \beta(n)x(n) + \epsilon(n+1) , \tag{1}$$

where

$$x(n) \quad = [x_1(n) \ \ldots \ x_7(n)]^T$$

$$\epsilon(n) \quad = [\epsilon_1(n) \ \ldots \ \epsilon_7(n)]^T \ , \ T \text{ stands for transposition}$$

$$
A \quad = \begin{bmatrix}
1-\alpha_1 & & & & & & a_{17} \\
 & 1-\alpha_2 & & & & & a_{27} \\
 & & 1-\alpha_3 & & \phi & & a_{37} \\
 & & \alpha_3 & 1-\alpha_4 & & & a_{47} \\
\phi & & & & 1-\alpha_5 & & a_{57} \\
 & & & & & 1-\alpha_5 & a_{67} \\
\alpha_1 & \alpha_2 & 0 & \alpha_4 & \alpha_5 & \alpha_6 & 1-\alpha_7
\end{bmatrix}
\tag{2}
$$

$$\alpha_7 \quad = \sum_{i=1}^{6} a_{i7} \tag{3}$$

$$
\beta(n) \quad = \begin{bmatrix}
-\beta_1(n) & & & & b_{17}(n) \\
 & & \ddots & \phi & \vdots \\
 & & & \ddots & \vdots \\
\phi & & \ddots & & \vdots \\
 & & & \ddots & b_{67}(n) \\
\beta_1(n) & \ldots & \ldots & \beta_6(n) & -\beta_7(n)
\end{bmatrix}
\tag{4}
$$

$$b_{i7}(n) \quad = \frac{a_{i7}}{\alpha_7} \beta_7(n) \quad \text{such that } \beta_7(n) = \sum_{i=1}^{6} b_{i7}(n) \ . \tag{5}$$

$$\epsilon_7(n+1) = \sum_{i=1}^{6} \epsilon_i(n+1) \tag{5}$$

where:

$x_i(n)$ = state of compartment i at time instant n, being the number of lymphocytes in compartment i expressed as a percentage of the total number in the system as measured by the presence of radioactive label.

α_i = deterministic part of transfer parameters (leaving i), a_{i7} is portion of α_7 going from 7 to i. α_i (and a_{i7}) scales the % activity in compartment i (or 7) at instant n (dimensionless).

$\beta_i(n)$ = stochastic part of transfer parameters (leaving i), $b_{i7}(n)$ is the portion of $\beta_7(n)$ going from 7 to i. $\{\beta_i(n)\}$ are i.i.d. and constitute multiplicative noise (dimensionless).

$e_i(n)$ = additive noise, added at the input and/or output of the compartment,

$\{\epsilon_i(n)\}$ are i.i.d. (dimensions of % activity).

$\{\beta_i(n)\}$ and $\{\epsilon_i(n)\}$ are independent. ϕ designates zero elements except where indicated.

The subscripts are as follows: 1 - bone marrow, 2 - spleen, 3 - other tissue, 4 - LN-a, 5 - LN-b, 6 - lungs, and 7 - blood.

It is assumed that there are no births and deaths, and no loss of label. Experimentally, the number of cells in any compartment is measured by the amount of radiolabel activity. The 7×7 matrix $\beta(n)$ is compartmental and is interpreted as the random coefficient of flow. $\epsilon(n)$ is a random 7-vector whose presence may be justified by changes in the same variables as for $\beta(n)$, but are due to intercompartmental stochasticity rather than intracompartmental randomness (as for $\beta(n)$). Also $\epsilon(n)$ may mimic the presence of different types of immunities and other physiological factors, and the random variation in different organisms within the same species.

Equation (1) can be viewed as a system of difference equations with both slope and intercept random or as a Markov chain generated by two stochastic processes (i.e., a Markov chain in a random environment), but such processes will not be discussed here. The second term on the right-hand side of Eq. (1) causes the structure to be that of a bilinear time series.

Literature is available on bilinear time series with one and two inputs (Nicholls and Quinn, 1982; Subba Rao and Gabr, 1984). A bilinear time series with two inputs has sometimes been called random coefficient autoregression. It is generally assumed that, in (1), A is a constant matrix, $\beta(n)$ is a matrix such that $E(\beta(n)) = 0$ and $E[\beta(n) \circ \beta(n)] = C_\beta$, and $\epsilon(n)$ is such that $E[\epsilon(n)] = 0$ and $E[\epsilon(n)\epsilon^T(n)] = G$. $\{\beta_i(n)\}$ and $\{\epsilon_i(n)\}$ are independent. $\circ$, the Kronecker product, is defined such that the ij-th block of A∘B is $[A \circ B]_{ij} = a_{ij}B$. Equation (1) may be used to estimate the mean and variance of the state. Results on second-order stability, second-order stationarity, their relationship, strict stationarity, and asymptotic properties of least squares and maximum likelihood estimators are given in Nicholls and Quinn (1982). Their biological interpretation that is relevant here follows.

If the variation in the number of recirculating lymphocytes in any compartment reaches an equilibrium which does not depend on the number of cells present initially in the compartment, the mean number of cells also reaches an equilibrium. This means that at equilibrium the distribution of these cells in that compartment has attained a unique constant mean and standard deviation. When the random perturbations in the recirculating lymphocyte pool are independent and identically distributed, the necessary and sufficient condition for the lymphocyte distribution in any compartment to have a constant mean and standard deviation at equilibrium is that the variation in the number of lymphocytes there be stable. Under these conditions the distribution is also asymptotically time-invariant. This implies that the number of recirculating lymphocytes can be characterized in terms of their population distributions in different compartments.

Smith and (late) Ford (1983) study the recirculating lymphocytes under conditions as close as possible to the physiological conditions. In the paper they report in effect three experiments: i) a survey of distribution of lymphocytes, ii) tempo of recirculation from blood to thoracic duct lymph, and iii) estimation of time taken for lymphocytes to cross high endothelial venules into LN. Only the one that is relevant to distribution is summarized here.

AO rats, adult male donors and adult female recipients were used. In vitro labeling of thoracic duct TDL was done with sodium-[^{51}Cr]-chromate, then passaged from blood to lymph in an intermediate rat to ensure using recirculating live lymphocytes. The results of the experiment are given in Table 1. Five rats were sacrificed at most of the 13 sample time points. After removal of the 13 relevant organs (viz., blood, lungs, spleen, liver, right and left popliteal LN, coeliac LN, superficial cervical LN, deep cervical LN, mesenteric LN, Peyer's patches, gut, and bone marrow) the results of scintillation counting were calculated as percentages of injected dose per organ and per gram of tissue. The results tabulated here are the means and standard deviations of activity per organ expressed as a percentage of the injected dose.

Soon after injection (1 minute) 40% of the injected dose is in the lung, 40% in the blood, and 13% in the liver and very little elsewhere. After 2 minutes the radioactivity in the lungs follows that in blood. Till about 1/2 hour it falls sharply in both and substantially in the liver. In the spleen, and all the LN and Peyer's patches the label steadily increases. Cell localization in gut and bone marrow seems flat from 1/2 hour on. At least in the spleen, LN, and Peyer's patches the rate of cell entry is apparently directly proportional to their concentration in the blood as suggested by experiments on isolated organs (Smith and Ford, 1983).

STATISTICAL ANALYSIS

Meaningful data interpretation is available in the form of state *means* and *standard deviations* which are obtained from Eq. (13) for parameter estimation. Estimates for the elements of A are obtained using the equation for means and those for the variances of the random variables from

$$P_d(n+1) = A_d P_d(n) + C_d P_e(n+1) \qquad (7)$$

where

$P_d(n)$ — a 7-vector diagonal of the variance-covariance matrix,

$$
A_d - \begin{bmatrix}
(1-\alpha_1)^2+\beta_1^2 & & & & & & a_{17}^2+b_{17}^2 \\
& \cdot & & \phi & & & \cdot \\
& & \cdot & & & & \cdot \\
& & & \cdot\ \cdot & & & \cdot \\
\phi & & & & \cdot & & \\
& & & & & \cdot & a_{67}^2+b_{67}^2 \\
\alpha_1^2+\beta_1^2 & \cdots & & & \alpha_6^2+\beta_6^2 & (1-\alpha_7{}^2+\beta_7^2)
\end{bmatrix}
$$

C_d — $\mathrm{diag}[\sigma_{\epsilon i}]$, $i - 1,\ldots,7$

and

$P_e(n) - \mathrm{diag}\{E[\epsilon(n)\epsilon^T(n)]\}$.

Here, $\sigma_{\epsilon i}$ refers to standard deviation, and ϕ denotes all zero elements except as indicated. Some important covariances, like those between blood and the other compartments, are not available in the experimental data, and Eq. (7) does not take them into account.

The weighted least squares cost function

$$
Qx^2 K(\cdot) - \frac{q}{nk} \sum_{i-1}^{k} \sum_{j-i}^{n} \frac{(x_{io}(j) - \hat{x}_i(j))^2}{\sigma_{io}^2(j)}
$$

is used. It is a function of the sample variance of the residual errors, so that effectively this is what is minimized to obtain the "best" parameter estimates. Being divided by variance from the data as weights, Q.(.) is a dimensionless ratio. For the first moments min $\chi^2 - 5.73$, while for the second moments min $Qx^2(C_\beta,G) -$ 116428.22. The estimated parameter values are tabulated in Table 2. The minimized χ^2-value is significant, which indicates that the residual errors are quite large. In other words, this preliminary model does not fit the data very well. However, according to their definitions given above with Eq. (1) α_i and a_{i7} provide, respectively, relative transfer rates of lymphocytes leaving the ith compartment and that portion of α_7 (leaving blood) which is associated with entering the ith compartment. While their accuracy is questioned here, their relative values have physical significance.

Here, A is an irreducible matrix. It is column stochastic and thus its spectral radius is 1. The seven compartments are not linearly independent because blood is the sum of the other compartments and overall it is a closed system.

$$H_0: \quad F_1 - F_2 - \ldots - F_7$$

$$H_1: \quad \text{Not all } F_1 \text{ are equal}$$

(8)

Friedman's two-way analysis of variance based on ranks was done. The null hypothesis concerns distributions (not means). For the first moments the test statistic $\hat{\chi}_R^2$ = 25.68 was obtained, which is significant ($\chi^2 - 16.81$, $k - 7$, $\alpha - 0.01$). Thus the errors in all the compartments do not have the same distributions. Wilcoxon and Wilcox multiple comparison procedure are used to do all pairwise comparisons. The distribution of errors in compartment 2 is significantly different from the distributions in compartments 1, 3, 4, 5, and 7 as seen from Table 3.

For the second moments also Friedman's analysis showed differences in distributions $\hat{\chi}_R^2 - 44.60 > \chi^2 - 16.81$, $k - 7$, $\alpha - 0.01$. Wilcoxon and Wilcox test gave three subsets of compartments with similar distributions:

$$\{(1,2,3,5),(1,2,5,6,7),(4)\}$$

It was shown above that for the estimated parameter values the model mean is marginally stable. Stability of the second moment is a function of the squares of the deterministic parts of the parameters and the variances of the parametric noise. Because of the parametric noise the model is a variable structure system (actually structurally unstable system) with random variation in the structure. As the estimated variances of the noises have large values in the model the distribution of recirculating lymphocytes is unstable. Comparing lymphocyte populations in any compartment there is a big difference in the estimated number of cells and the experimental quantity. For the mean, compartment 2 (spleen) appears to be the main source of error, while for the variance (standard deviation) compartments 3 (LT) and 4 (LN-a) may be the main cause.

Apparently, the model can be significantly improved by appropriate covariance terms, appropriately correlated parametric noise sequences and/or feedback of appropriate states to generate the diffusion coefficients - all of which would require further data.

ACKNOWLEDGEMENTS

This research was supported by NSF Grant No. EE T-8618062.

Table 2.   Estimated Parameters Values for the Discrete
Model Weighted Least Squares Estimation

Deterministic Parts of Multiplicative Parameters

α_1	.11948E-01	a_{37}	.35198E-01	α_6	.38460E+00
a_{17}	.48594E-01	α_4	.33771E-02	a_{67}	.39131E+00
α_2	.78494E-02	a_{47}	.34741E-02	α_7	.49617E+00
a_{27}	.65323E-01	α_5	.41243E-03		
α_3	.15659E-02	a_{57}	.36640E-02		

Standard Deviations of the Multiplicative Noises

$\sigma_{\beta 1}$	.10245E-02	$\sigma_{\beta 4}$	.15762E-03	$\sigma_{\beta 6}$	.93364E+00
$\sigma_{\beta 2}$	.99294E+00	$\sigma_{\beta 5}$	.10000E+01	$\sigma_{\beta 7}$	.29113E+00
$\sigma_{\beta 3}$	.99673E+00				

Standard Deviations of the Additive Noise

$\sigma_{\epsilon 1}$	.23895E-01	$\sigma_{\epsilon 4}$	.56450E-05	$\sigma_{\epsilon 6}$	.37201E-06
$\sigma_{\epsilon 2}$	.19190E-05	$\sigma_{\epsilon 5}$	.74743E-01	$\sigma_{\epsilon 7}$	.23510E+01
$\sigma_{\epsilon 3}$	.22762E+01				

With regards to the second moments, the matrix $(A \circ A + C_\beta)$ is a 49×49 sparse matrix, and the usual method for finding eigenvalues is not efficient. Ranges of eigenvalues were estimated using Gershgorin's Circle Theorem (Lancaster, 1969) and the fact that the spectral radius of a matrix is always greater than or equal to the largest diagonal element. The spectral radius is greater than 1. (The eigenvalues lie in the interval $[-6.27, 8.5]$ and the spectral radius > 1.9992).

The implication is of second-order instability and second-order nonstationarity (Nicholls and Quinn, 1982). This being so and the minimized χ^2 being significant at $\alpha = 0.01$, further analysis was done to check into the goodness of fit of the compartments of the model. Since one of the conditions for existence and convergence of moments, $\rho(A \circ A + C_\beta) < 1$, is the same as the condition for stability which is violated, the second moment diverges (Feigin and Tweedie, 1985). There is a possibility of this happening because the variances are estimated by using only the diagonal of the $E[x(n)x^T(n)]$ matrix, covariances not being available in the data. By doing so the number of equations used in the estimation of the parameters is reduced and in turn the number of constraints on the parameter is also reduced thus leading to "misestimation."

Since the exact distributions are not known the following null and alternative hypotheses were tested on the residual errors between the model output and the experimental data:

REFERENCES

1. Bell, G.I., A.S. Perelson, and G.H. Pimbley, Jr. (1978), "Theoretical Immunology," *Immunology Series* (New York: Marcel Dekker, Vol. 8
2. DeLisi, C. (1983), "Mathematical Modeling in Immunology," *Ann. Rev. Biophys. Bioeng.*, 12, 117-138.
3. DeBoer, R.J., P. Hogeweg, et al. (1985), "Macrophage T Lymphocyte Interactions in the Anti-Tumor Immune Response," *J. Immunology*, 134, 1-11.
4. DeSousa, M. (1981), *Lymphocyte Circulation: Experimental & Clinical Aspects*, (New York: John Wiley & Sons).
5. Feigin, P.D., and R.L. Tweedie (1985), "Random Coefficient Autoregressive Processes: A Markov Chain Analysis of Stationarity and Finiteness of Moments," *J. Time Series Anal.*, 6(1), 1-14.
6. Fung, Y.C. (1984), *Biodynamics: Circulation*, (New York: Springer-Verlag).
7. Lancaster, P. (1969), *Theory of Matrices*, (New York: Academic Press).
8. Lipster, R.S., and A.N. Shiryayev (1977), *Statistics Processes I - General Theory* (New York: Springer-Verlag).
9. Lipster, R.S., and A.N. Shiryayev (1978), *Statistics Processes II - Applications* (New York: Springer-Verlag).
10. Marchuk, G.I. (1983), *Mathematical Models in Immunology* Translation, Series in Mathematics and Engineering (New York: Optimization Software).
11. Marchuk, G.I., and L.N. Belykh (eds.) (1983), "Mathematical Modeling in Immunology and Medicine," *Proceedings of the IFIP-TC 7 Working Conference on Mathematical Modeling in Immunology & Medicine, Moscow, USSR, 5-11 July, 1982* (Amsterdam: North-Holland Publishing Co.).
12. Marchuk, G.I., L.N. Belykh, and S.M. Zuev (1985), "Mathematical Modelling of Infectious Diseases," *Problems of Computational Mathematics and Mathematical Modelling, Advances in Science and Technology in U.S.S.R. (Mathematics and Mechanics Series)*, Eds. G.I. Marchuk and V.P. Dymnikov (Moscow: Mir Publishers, 223-240.
13. Merrill, S.J. (1980), "Mathematical Models of Humoral Immune Response," *Modeling and Differential Equations in Biology, (Lecture Notes in Pure & Applied Mathematics)*, Ed. T.A. Burton (New York: Marcel Dekker), 13-50.
14. Mohler, R.R. (1973), *Bilinear Control Processes* (New York: Academic Press).
15. Mohler, R.R. (1987), "Foundations of Immune Control and Cancer," *Recent Advances in System Science*, Ed. A.V. Balakrishnan (New York: Optimization Software, Inc.), (to appear).
16. Mohler, R.R., C. Bruni, and A. Gandolfi (1980), "A Systems Approach to Immunology," *Proc. IEEE* 68(8), 964-990.
17. Nicholls, D.F., and B.G. Quinn (1982), "Random Coefficient Autoregressive Models: An Introduction," *Lecture Notes in Statistics* (New York: Springer-Verlag), Vol. 11.
18. Roitt, I. (1974), *Essential Immunology* (Oxford and London: Blackwell Scientific), 162.
19. Rubinow, S.I. (1975), *Introduction to Mathematical Biology* (New York: John Wiley and Sons).
20. Smith, M.E., and W.L. Ford (1983), "The Recirculating Lymphocyte Pool of the Rat: A Systematic Description of the Migratory Behavior of Circulating Lymphocytes," *Immunol.* 49, 83-94.
21. Subba Rao, T., and M.M. Gabr (1984), "An Introduction to Bispectral Analysis and Bilinear Time Series Models," *Lecture Notes in Statistics* (New York: Springer-Verlag), Vol. 24.

Table 1. Original Data from Smith & Ford (1983)

(mean + standard deviation)

% radioactive label (relative to injected dose) per organ

Time:	0	1	2	5	10	30	60	150	360	540	720	900	1080	1440
superficial cervical LN	0.00	0.18 +.06	0.24 +.11	1.02 +.68	2.20 +1.01	5.15 +1.18	7.46 +.76	7.71 +1.91	8.43 +1.72	8.97 +1.72	9.40 +1.13	8.88 +.43	8.75 +.99	7.83 +1.53
deep cervical LN	0.00	0.06 +.02	0.09 +.04	0.49 +.30	1.11 +.45	2.17 +.47	3.55 +.40	3.63 +.95	4.35 +.40	4.32 +.64	4.99 +.74	4.62 +.55	4.29 +.28	3.97 +.54
coeliac LN	0.00	ns	ns	0.09 +.07	0.15 +.06	0.41 +.15	0.60 +.13	0.60 +.13	0.84 +.17	0.71 +.18	0.94 +.16	0.84 +.14	0.68 +.17	0.81 +.13
mesenteric LN	0.00	0.14 +.05	0.17 +.07	0.86 +.65	1.66 +.65	4.91 +1.40	8.18 +1.43	9.32 +1.40	10.79 +1.32	12.65 +3.80	15.97 +3.25	16.34 +1.55	17.27 +2.48	15.62 +1.71
Peyer's patches	0.00		0.19	0.85 +.70	1.58 +.69	3.97 +1.07	6.68 +.96	5.73 +.50	6.16 +1.06	6.09 +.90	7.46 +1.53	6.68 +1.23	6.44 +.99	6.39 +1.27
small intestine	0.00	0.90 +.21	0.87 +.18	1.29 +.69	1.35 +.39	1.20 +.45	0.81 +.78	1.69 +.66	2.05 +.99	2.65 +1.35	1.99 +1.23	2.56 +1.47	1.54 +.51	1.72 +1.02
spleen	0.00	2.29 +1.79	5.31 +1.68	10.56 +2.64	19.51 +6.04	38.92 +5.45	39.41 +5.58	32.31 +3.42	21.32 +2.97	17.90 +3.72	19.14 +3.33	14.86 +3.22	15.48 +1.92	15.48 +.78
lung	0.00	35.50 +6.14	40.08 +6.13	27.63 +6.73	22.97 +4.87	8.46 +1.25	2.50 +.60	2.73 +.51	2.33 +.20	1.85 +.26	1.66 +.33	2.02 +.36	1.86 +.41	2.03 +.38
liver	0.00	12.84 +2.76	12.13 +2.82	16.24 +6.16	11.24 +2.70	6.22 +.83	3.92 +.83	3.66 +.39	3.66 +.58	4.04 +.39	4.11 +.45	4.56 +1.09	4.24 +.32	5.26 +.51
bone/tibia	0.00	1.15 +.21	1.15 +.21	1.36 +.49	1.61 +.44	1.81 +.69	2.26 +.50	2.23 +1.29	1.33 +.38	1.24 +.55	2.17 +1.22	0.72 +.34	1.01 +.50	0.73 +.15
blood/10ml	100.	41.47 +4.94	36.59 +1.80	38.91 +4.80	29.58 +5.34	11.64 +2.93	3.34 +1.15	4.06 +.88	3.25 +.34	3.06 +.52	2.41 +.56	3.13 +.92	2.46 +.58	3.04 +.65
R. popliteal LN	0.00				0.24 +.07	0.63 +.16	0.73 +.09	0.72 +.13	0.77 +.10	0.78 +.20	0.78 +.06	0.84 +.17	0.73 +.04	0.75 +.14
L. popliteal LN	0.00				0.06 +.02	0.16 +.06	0.21 +.04	0.23 +.06	0.22 +.06	0.28 +.08	0.33 +.06	0.28 +.05	0.25 +.02	0.23 +.03

Table 3. Statistical Analyses: Discrete-Time Model

First Moments Weighted Least-Squares Estimation

1(a): Friedman Test

Compartment #	Sum of Ranks		
1	59	3	(64)
2	24	5	(63)
3	64	7	(62)
4	58	1	(59)
5	63	4	(58)
6	34	6	(34)
7	62		

$\hat{X}^2_R = 25.68**$

1(b): Wilcoxon and Wilcox Text

From tables: $D_{WW;a} = \begin{bmatrix} 32.5 & a = 0.05 \\ 38.0 & a = 0.01 \end{bmatrix}$

	5 (63)	7 (62)	1 (59)	4 (58)	6 (34)	2 (24)
3 (64)	1	2	5	6	30	40**
5 (63)		1	4	5	29	39**
7 (62)			3	4	28	38**
1 (59)				1	25	35*
4 (58)					24	34*
6 (34)						10

Second Moments Weighted Least-Squares Estimation

Compartment #	Sum of Ranks		
1	50	3	(82)
2	72	2	(72)
3	82	5	(58)
4	17	1	(50)
5	58	6	(44)
6	44	7	(41)
7	41		

	3 (72)	6 (58)	1 (50)	6 (44)	7 (41)	4 (17)
3 (82)	10	24	32	38**	41**	65**
2 (72)		14	22	28	31	55**
5 (58)			8	14	17	41**
1 (50)				6	9	33*
6 (44)					3	27
7 (41)						24

$\hat{X}^2_R = 45.33**$

* indicates significance at a = 0.05.
** indicates significance at a = 0.01.

Acta Applicandae Mathematicae 14 (1989), 155–166.
© 1989 *by IIASA*.

Recent Progress in 3-D Computer Simulation
of Tumor Growth and Treatment

Werner Düchting

Fachbereich Elektrotechnik
Universität Siegen
Hölderlinstr. 3, D-5900 Siegen
West Germany

AMS Subject Classification (1980): 92A07
Key words: systems analysis, spatial simulation

1. Introduction

During the last decade many biological experiments were made to get
more insight into the biological behaviour of cells and about cancer
(1). In parallel with these experiments mathematicians were stimulated to apply their science to biological questions ((2)-(5)). That
means they try to describe the time behaviour of cell interactions by
highly nonlinear differential equations of high order. The solutions
gained by calcultion or simulation always show the response curve as
a function of time. In many cases, however, the influence of feedback
loops and the description of the spatial structure of growing cell
systems are neglected. Therefore, our task is to investigate how
methods of systems analysis control theory and computer science can
be applied to cancer problems to answer the question for normal and
malignant cell growth in time and space.

2. The cancer problem

In this century several hypotheses of the development of cancer have been the subject of investigation: virus theory (1910 by Rous), mutation theory (1914 by Boveri), and metabolism theory (1926 by Warburg). Recent research of the genetic alterations that cause cells to become malignant ones has focussed on oncogenes. In this rapidly moving research area studies have revealed that dominant cellular genes called "proto-oncogenes" are activated by tumor viruses, gene amplification, gene translocation and genetic mutation. In spite of this progress the main question how genes and the growth of normal and malignant cells are regulated still remains open.

Instantaneously it is not possible to construct computer models describing the mechanims of gene control because of the lack of experimental data. Therefore, we have to confine ourselves to the cellular level. Even at this level and at present we are unable to consider all characteristic features of malignant tumors (uncontrolled proliferation, invasion in adjacent tissue, metastases, ability to evade immune surveillance). Thus, several oversimplifying assumptions have to be stated as for instance a cubic shape of a cell or the neglection of heterogenity, immunologic reactions and formation of metastases. With these underlying assumptions we will try to construct different types of models (6) which enable to simulate the time behaviour of disturbed cell-growth control circuits, to predict spatial tumor growth (2-D, 3-D) and to simulate different kinds of cancer treatment (surgery, radiation- and chemotherapy).

3. Design philosophy

When constructing a computer model of a complex biological system, it is necessary to design a modular concept. In this case of modelling tumor growth it means to design modular structured subsystems (7).

(i) We need feedback control models which describe the cell division of normal and tumor cells at a cellular level including experimentally gained data e.g. of cell-cycle phase durations.

(ii) Heuristic cell-production and interaction rules are required describing the cell-to-cell communication. For instance one rule of the catalogue may say: All tumor cells residing at a distance larger than 100 μm from the capillaries after the next division step will enter the resting phase G0.

(iii) Cell movement is described by transport equations (diffusion-, Poisson-equation), that means we have to introduce gradients for pressure and metabolic compounds into the model.

(iv) To represent 2-D and 3-D simulation results computergraphics software packages are necessary.

These statements, rules and equations have been transformed into algorithms, and subsequently subprogramme packages have been written in FORTRAN IV. To start the simulation programme packages the follo- wing input data have to be fed into the computer (VAX 730): Notations about the character of a cell (normal, malignant), cell-cycle phase durations, cell-loss rates, initial configuration of normal tissue and of tumor cells and distinguished data about the kind of the planned tumor treatment.

4. Previous approaches

Only very few contributions consider biomedical problems from the viewpoint of control theory (8,9).

Feedback models describing the disturbed time behaviour of red-blood cells: Our own approach developing closed-loop control circuits for tumor growth started in 1968 (10). At that time the subject of consi- deration was focussed on stability conditions and on the interpreta- tion of cancer as an unstable closed-loop control circuit. Step by step the dynamic behaviour of cell-renewal control loops (Fig. 1) was investigated. Blockoriented simulation languages were used for simu- lating the macromodels. As a result the number of cells as a func- tion of time was plotted (Fig. 2).

2-D spatial models describing malignant cell growth and simulation of tumor treatement: Then oncologists advised us to consider not only the time, but also the spatial behaviour of tumor growth. Figure 3 illustrates the two-dimensional growth of tumor cells (*) in a normal

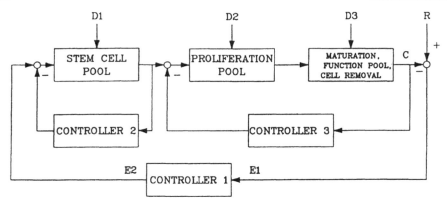

R: Required tissue oxygen (desired number of erythrocytes)

C: Number of red blood cells (erythrocytes)

E2: Production of the erythropoietin hormone

D1,D2,D3: Disturbances

Fig. 1: Multi-loop control circuit of erythropoiesis

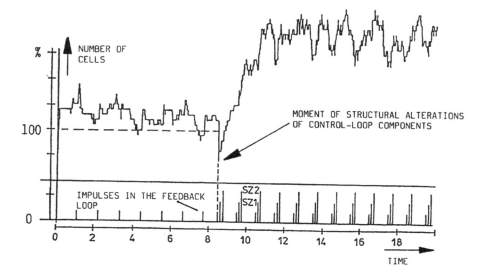

Fig. 2: Number of cells as a function of time indicating a cancerogeneous
behaviour (simulation result)

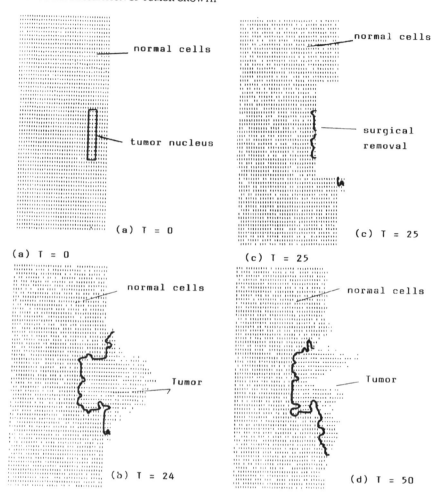

Fig. 3: Simulation of two-dimensional tumor growth and surgical treatment

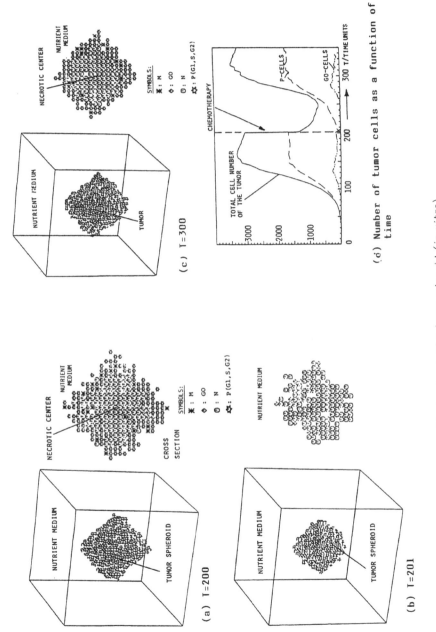

(a) T=200

(b) T=201

(c) T=300

(d) Number of tumor cells as a function of time

Fig. 4: Simulation of a chemotherapeutic treatment of a tumor spheroid (in vitro)

tissue (1) and a partial surgical removal of the tumor at T=25 units of time. It is evident that under this condition a total regression of the tumor cannot be expected. It should be noted that this model only allows to distinguish between normal cells and tumor cells (11).

<u>3-D spatial models describing tumor growth in vitro and simulation of chemotherapy</u>: By introducing distinguished cell-cycle phases (G1, G2, M, GO, N) we were able to simulate the 3-D growth of a single dividing tumor cell inoculated into the center of the cell space of a nutrient medium at the beginning of the simulation run (12). Figure 4 shows the result of a test run simulating a chemotherapeutic treatment of a tumor spheroid in vitro. In this case the assumption was made that all proliferating tumor cells (i.e. the outside rim) in Figure 4(b) were killed by a cytotixic drug at T=201 time units. By this treatment the resting GO-tumor cells are now residing in a very short distance to the nutrient medium, that means they can be recruited into the cell cycle again, and tumor growth continues (Fig. 4(c)). The computer simulation enables the determination of the optimum time when to apply the cytotoxic drug for a second time. This value is given by the minimum of the GO-tumor cells.

5. Spatial models of in-vivo tumor growth

After simulting in-vitro tumor growth the attempt was made to substitute the nutrient medium by static blood vessels. However, very soon it was clear that a more realistic structure of capillaries was desirable for simulating in-vivo tumor growth. Therefore, Vogelsaenger (13) constructed a model which described the spatial formation of capillaries as a regulation process controlled by the request for energy (O_2, glucose) of each cell of an organ in evolution. A comparison between Figure 5 and Figure 6 shows that for the cortex of a rat the simulation result is highly similar to the experimental result received by Bär (14). The formation of capillaries is the presumption for modeling the spatial evolution of a normal brain segment which was performed in (13).

RIM

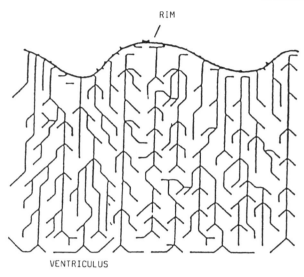

VENTRICULUS

Fig. 5 : Capillary network in the cortex (simulation result)

RIM

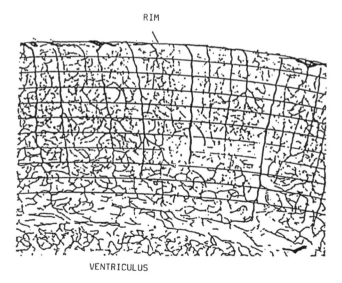

VENTRICULUS

Fig. 6 : Vascularization of the cortex (14)

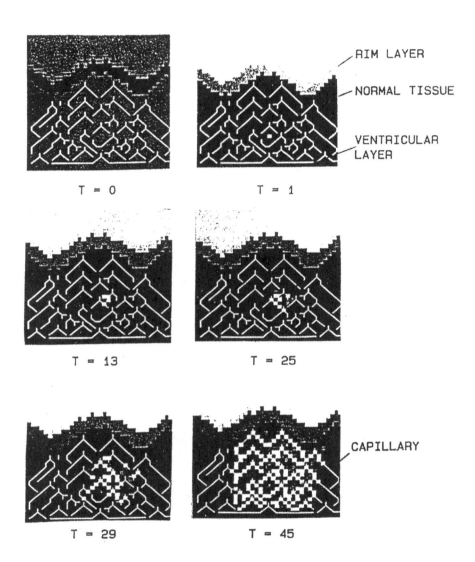

Fig. 7: Spread of tumor cells in the cortex (simulation result)

Now the assumption is made that a single tumor cell is arbitrarily placed in the tissue of the cortex at T=1 unit of time. If this tumor cell resides close to a capillary it will divide and move in accordance with the cell production rules (Fig. 7). Further, tumor growth is possible only because tumor cells produce a substance which is called tumor-angiogenesis factor (TAF). This factor stimulates nearby blood vessels to send out new capillaries (Fig. 8) which grow towards the tumor. penetrate it and lead to further rapid tumor growth.

Future work will be concentrated on the investigation of predicting tumor growth with this model, simulating cancer treatment (surgery, radio- and chemotherapy) and optimizing tumor treatment prior to clinical therapy.

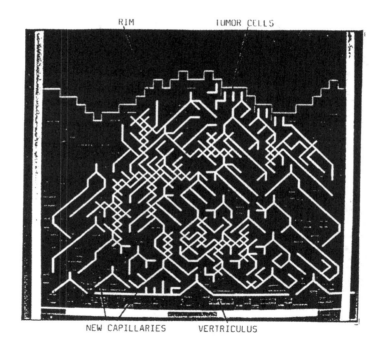

Fig. 8: Simulation of the tumor angiogenesis effect: Formation of new capillaries
        (T = 120 units of time)

6. <u>Outlook</u>

From the catalogue of unsolved problems in the area of cancer
modeling I think there are three promising avenues of future
work:

- Consideration of facts which had to be neglected so far (formation of metastases, immunologic reactions, drug resistance, heterogenity, side effects).
- Generation of a more realistic initial configuration of a   tumor
  by combining CT-pictures (Computer tomography) with   predictive
  models describing tumor growth and last not least
- Optimization of distinguished methods and schedules of cancer
  treatment.

7. References

(1) Tannock, I.F. and Hill, R.P. (edt.), The Basic Science of Onco-
     logy, Pergamon Press, New York 1987.

(2) Marchuk, G.I. and Belykh, L.N. (eds.), Mathematical Modeling in
     Immunology and Medicine, North Holland, Amsterdam 1983.

(3) Hoffman, G.W. and Hraba, T. (eds.), Immunology and Epidemiology,
     Springer-Verlag, Berlin 1986.

(4) Meinhard, H., Models of Biological Pattern Formation, Academic
     Press, London 1982.

(5) Thompson, J.R. and Brown, B.W. (eds.), Cancer Modeling, Marcel
     Dekker, New York, 1987.

(6) Düchting, W., Computer Simulation in Cancer Research, in "Advan-
     ced Simulation in Biomedicine," edited by D.P.F. Möller, Sprin-
     ger-Verlag, Berlin, to appear in 1988.

(7) Düchting, W. and Vogelsaenger, Th., Aspects of Modelling and
     Simulating Tumor Growth and Treatment, J. Cancer Res. Clin.
     Oncol. 105(1983): 1-12.

(8) Swan, G.W., Applications of Optimal Control Theory in Biomedi-
     cine, Marcel Dekker Inc., New York 1984.

(9) Mohler, R.R. and Hsu, C.S., Systems Compartmentation in Immuno-
     logical Modeling, in "Systems Theory in Immunology" edited by C.
     Bruni, G. Doria, G. Koch and R. Strom, Springer-Verlag, Berlin
     1979, 165-174.

(10) Düchting, W. Krebs, ein instabiler Regelkreis, Versuch einer
     Systemanalyse, Kybernetik, 5. Band, 2.Heft (1968): 70-77.

(11) Düchting, W. and Dehl, G., Spatial Structure of Tumor Growth: A
     simulation Study, IEEE Transactions on Systems, Man and Cyberne-
     tics SMC-10, No. 6 (1980): 292-296.

(12) Düchting, W. and Vogelsaenger, Th., Three-Dimensional Pattern
     Generation applied to Spheroidal Tumor Growth in a Nutrient
     Medium, Int. J. Bio-Medical Computing 12 (1981): 377-392.

(13) Vogelsaenger, Th., Modellbildung und Simulation von Regelungs-
     mechanismen wachsender Blutgefäßstrukturen in normalen Geweben
     und malignen Tumoren, Dissertation Siegen, 1986.

(14) Bär, Th., Patterns of Vascularization in the Developing Cere-
     bral-Cortex, CIBA Found. Symp. 100(1983): 20-36.

Acta Applicandae Mathematicae **14** (1989), 167–178.

Mathematical Simulation of the Immunomodulating Role of Energy Metabolism in Support of Synergism and Antagonism of the Compartments of an Immune System

Vasilij M. Janenko and Constantin L. Atoev

V.M. Glushkov Institute of Cybernetics

Academy of Sciences of the Ukrainian SSR

252207 Kiev, USSR

AMS Subject Classification (1980): 92A07

Key words: immune response, metabolism, simulation

1. Introduction

The idea that pathology can be considered as the unity of two opposite processes - the degradation of intracellular structures which result in energy metabolism disturbance and the realization of some adaptive control mechanism, formed in the process of evolution, which support the values of vital parameters optimal for the present state of the organism - is one of the axioms of contemporary theoretical biology. When considering a pathological process from this viewpoint we may come to the conclusion that the problem of therapy is reduced to the definition of a set of actions which activate natural prevention mechanisms [3] of the organism and create an optimal condition for their realization. Mutual regulation of the processes of synthesis and consumption of energy in a cell occupy central places in the understanding of the activity of such molecular regulation mechanisms responsible for the adequate energy support of physiological functions of the organism. Without the study of this regulation we cannot understand and estimate properly the role of metabolic changes in a cell which accompany the various pathological processes and therefore we cannot choose properly a complex of therapeutic actions. The aim of the given paper is to investigate the role of energy metabolism in support of the balance of some branches of the immune system.

2. A Mathematical Model

Energy factors play an important part in the immune response since the processes of proliferation and differentiation, on the one hand, and the processes of recognition and destruction of viruses and affected cells, on the other hand, are energodependent processes [3].

The role of energy factors in regulation of the immune response may be considered from the viewpoint of the energetics of the immune system as well as from the viewpoint of the energetics of the damaged organ [9]. A system of cyclic nucleotides is an important element through which the immunomodulating role of energy metabolism in the regulation of intracellular processes is realized [4]. In the result of antigen - antibody interaction (AAI) their receptors acquire enzymatic activity and the membrane permeability for Ca^{2+} - ions increases. The growth of Ca^{2+} in a cell results, on the one hand, in the change of cAMP concentration and, on the other hand, activates Ca^{2+} - dependent ATPase of membrane and changes the rates of the processes of synthesis and consumption of energy in a cell [11].

A mathematical model for the investigation of the role of energetics in the immune response therefore must include blocks of the immune system, of energy metabolism and Ca^{2+} exchange. To construct the model we use the following assumptions:

(1) Idiotype-antiideotypic interactions of subpopulations of lymphocytes of the immune network are substantiated by hypotheses of Mikhalevich et al. [12] (1986) and in the given paper are represented in the integral form.

(2) Mechanisms of anti-tumoral protection realized in the interaction between the immune system and oncogens have a time hierarchy. This permits of distinguishing a group of slow variables which include a size of the tumor - X; effector population of lymphocytes - Y_1, incorporating killer and helper lymphocytes; interleukin - 2 - Y_2; suppressor population of lymphocytes - Y_3, incorporating specific and non-specific lymphocytes.

(3) The immune system is one of the regulators of angiogenesis. Suppressor lymphocytes can affect either the processes of tumor destruction or the processes which stimulate its growth. The first mechanism is connected with inhibition of effectors, the second - with inhibition of the action of mast cells as activators of the angiogenesis [7].

(4) Interleukin-2 which can have variously directed effect on subpopulations of immunocompetent cells is one of the most powerful factors influencing the immune system - oncogen interaction (ISOS) [4].

(5) The cAMP system plays an important part in the regulation of pathogenesis since the growth of AMP leads to a termination or weakening of the allergic reaction, and its fall is caused by the intensive AAI. There exists an antagonism between the action of cAMP and interleukin-2 since cAMP inhibits an cytoxic effect of natural killers and interleukin-2 activates it [5].

(6) After AAI the change of membrane permeability for Ca^{2+} occurs. The increase of Ca^{2+} in a cell results in the strengthening of bronchospasm since mediators are released from mast cells as Ca^{2+} enters them. Verapamil and nifedipin, blockers of slow calcium channels, allay the bronchospasm caused by histamine [6]. Histamine has an active effect on the development of allergic reactions.

(7) A brochial asthma attack (BAA) appears in the following form: change of the tension of small and smallest bronchia, bronchospasm and bronchial obstruction. Mast cells in the bronchial asthma attack play the part of cell-targets. Mediators which are released in the process of reaction from the mast cells are: histamine, heparin and a number of other substances [1].

(8) A contractile system of airways smooth muscle is Ca^{2+} dependent and is regulated, at least partially, by the entering of Ca^{2+} into a muscle cell [11]. The influx of Ca^{2+} into smooth muscle is proportional to the level of released histamine and the amount of evolved histamine is proportional to the level of Ca^{2+} in mast cells. The flow of Ca^{2+} into the mast cells is proportional to the rate of AAI on the cell surface. The growth of Ca^{2+} results in the growth of cAMP and this decreases the activity of effector cells.

The mathematical model will have the following form

$$\frac{dX}{dY} = k_1 X U_1(Y_3) - k_2 X - k_3 X Y_1 U_2(Y_3),$$

$$\frac{dY_1}{dt} = k_4 X - k_5 Y_1 - m k_3 X Y_1 U_2(Y_3) + U_3(Y_1, Y_2, C_2) k_6,$$

$$\frac{dY_2}{dt} = k_7 U_4(Y_1, Y_2) - k_8 Y_2,$$

$$\frac{dY_3}{dt} = k_9 U_5(Y_2, Y_3) - k_{10} Y_3,$$

$$\frac{dC_1}{dt} = b_1 k_{11} F_1 + k_{12} X Y_1 U_2(Y_3) - k_{13} U_6(C_1, F_1) - k_{14} U_7(C_1), \qquad (5)$$

$$\frac{dF_1}{dt} = k_{13} U_6(C_1, F_1) - k_{11} F_1,$$

$$\frac{dG}{dt} = k_{15} F_1 - k_{16} G,$$

$$\frac{dC_2}{dt} = k_{17} G - k_{18} U_8(C_2),$$

$$\frac{dF_2}{dt} = k_{20} U_9(F_2, C_3) - k_{21} F_2,$$

$$\frac{dC_3}{dt} = k_{19} G - k_{20} U_9(F_2, C_3) + b_2 k_{21} F_2 - k_{22} U_{10}(C_3),$$

where

$$U_1 = 1 - \frac{S_1 Y_3}{S_2 + Y_3}; \quad U_2 = 1 - \frac{N_1 Y_3}{N_2 + Y_3}, \quad U_3 = \frac{Y_1 Y_2}{S_3 + Y_1} \left(1 - \frac{S_4 C_2}{S_5 + C_2}\right),$$

$$U_4 = \frac{Y_1 Y_2}{S_6 + Y_2}, \quad U_5 = \frac{Y_2 Y_3}{S_7 + Y_3}, \quad U_6 = \frac{C_1(1 - F_1)}{N_3 + C_1}, \quad U_7 = \frac{C_1}{N_4 + C_1},$$

$$U_8 = \frac{C_2}{N_5 + C_2}, \quad U_9 = \frac{C_3(1 - F_2)}{N_6 + C_3}, \quad U_{10} = \frac{C_3(1 - F_2)}{N_7 + C_3}.$$

C_1, C_2 and C_3 are levels of Ca^{2+} in mast cells, effectors and the contractile system, respectively; F_1 and F_2 are portions of activated histamine pools and bounded tropomyosine by which the spasm intensity is estimated, respectively; G is the level of histamine; X is the level of antigen; $k_i (i=1,\ldots,22)$, N_j, $S_{j_{2+}} (j=1,\ldots,7)$ are model parameters; $b_r (r=1,2)$ are the number of Ca^{2+} ions bounded with one molecule of protein.

3. ISOI Investigation

The problem is formulated in the following manner: (a) to specify the character of interaction between the tumor growth and the lymphocyte activity; (b) to estimate a possible contribution of suppressors and effectors into ISOI; (c) to estimate the efficiency of indirect influence of interleukin-2 on the tumor growth and to determine the character of interrelation of this influence and the kinetics of the product of this mediator as the tumor grows; (d) to determine the conditions for oscillatory states in ISOI which determine chronic forms of diseases.

When investigating a model (1) on the basis of time hierarchy we will assume that the fast variables of the system succeeded in attaining their stationary states. Then the interaction is determined only by the variables X, Y_1, Y_2, Y_3. We consider the following variants of interactions: 1) in the complete absence of effect of Y_3 when k_9, k_{10}, S_1, N_1, equal zero, $U_3 = X_2$, $U_4 = X$; 2) accounting for the effect of suppressors but in the complete absence of Y_2, when k_7, k_8, S_1, N_1 equal zero, $U_5 = Y_1$, $k_4 = k_{40} (N - Y_3)$; 3) accounting for a suppressor effect on Y_1 as well as for tumor growth.

The investigation of the mathematical models of *ISOI* has shown the following. Lymphocytes can affect either the processes of tumor destruction or the processes stimulating its growth. The first mechanism leads even in the complete absence of suppression to tumor growth if $k_1 > k_2$. The suppressor population in this case only worsens the situation. The second mechanism is related to the fact that with small tumors when the activity of the suppressor population is low then for a substantial inhibition of anti-tumor action of Y_1, suppressors should simultaneously inhibit the processes which encourage the tumor growth, e.g. angiogenesis. Thus it is believed that if the tumor is small, the action of suppressors and effectors is additive, but at later stages of interaction they come into antagonism.

The stationary level of Y_2 has an important value for the dynamics of the tumor growth. There are conditions under which a further increase of interleukin-2 will favor the progressive growth of tumor. This, on the one hand, theoretically substantiates experimental data (see Figure 3), and on the other hand, defines the ways for quantitative determination of optimal doses of locally introduced interleukin-2 in each specific case of the tumor growth for obtaining the anti-tumor effect. In the process of ISOI, oscillation modes may appear, and this may cause chronic forms of the disease. The results of computer experiments are presented at Figure 1 (- - is the model curve, --- is the experimental curve (see [7])).

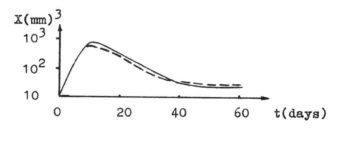

Figure 1

4. The Investigation of BAA

The problem is formulated in the following manner: (a) to investigate the character of the effect of calcium channels blockers on the intensity of allergic reactions; (b) to investigate the dependence of the immune response on the intensity of the pump system of the cell (Ca^{2+} ATPase); (c) to investigate the dependence of the immune response on the rate of reproduction of effector cells as reaction to antigen introduction; (d) to investigate the dependence of immune response on parameters characterizing the suppressor action.

Mathematically the problem is formulated in the following form: to define functions $F_{max}(k_4)$, $F_{max}(k_{12})$, $F_{max}(k_{14})$, $F_{max}(k_{15})$, where F_{max} is the maximal value of the function F_2 on the interval $[0, T]$; k_4, k_{12}, k_{14}, k_{15} are constants characterizing, respectively, the sensitivity of a system of antigen recognition, the rate of Ca^{2+} influx into the cell, the rate of disconnection of the complexes of Ca^{2+} with intracellular calcium- bounding proteins (actomyosine complexes), the rate of calcium pump.

Results of the modelling obtained for variations of the given constants are shown in Figure 2.

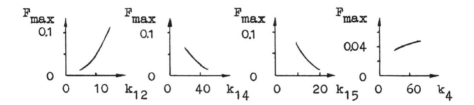

Figure 2

The increase of k_4 results in the growth of the amplitude of the spasm F_{max}, and this is connected with the growth of the AAI which is caused by the growth of Y_1. The increase of parameters k_{14} and k_{15}, on the contrary, lowers the spasm amplitude. The decrease of parameter k_{12} stimulates the action of calcium-channel blockers.

Figure 3 shows different models of the development of bronchospasm which may be obtained with the help of the model (1); contraction with constant tonic constituent which is characteristic for smooth muscles; (a) rythmic contraction of smooth muscles (b).

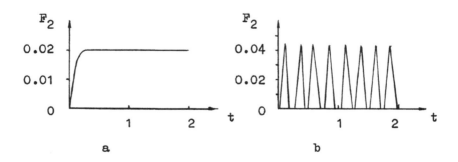

Figure 3

Figure 4 shows the results of simulation for variations of parameters N_j, S_j ($j=1,2$) which characterize the action of suppressors.

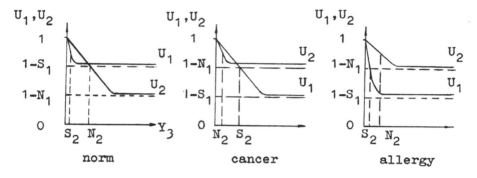

Figure 4

When the conditions $S_2 < N_2$, $S_1 < N_1$ are satisfied, we have a model characterizing the norm; at $S_2 > N_2$, $S_1 > N_1$ the model of immune response characterizes a predisposition to ongologic diseases; at $S_2 < N_2$, $S_1 > N_1$ the population of Y_3 appears to be suppressed and this may result in the development of allergic diseases.

5. The Investigation of the Effect of Energy Factors on AAI

A model of the immune response which may be used for the investigation of the role of energetics in providing the balance between individual chains of the immune system was suggested in [8]. We supplement this model with the regulation circuit which relates the energetics to defensive functions of the organism.

Eliminate from the model (7.77) suggested in ([8], p. 304) the variables which describe the clone hemopoiesis cells and a negative clone of lymphopoieses cells and retain the following variables: m_0, m_1, m_3, m_4 respectively characterizing the rates of reproduction of cells precursors, a positive clone of lymphopoiesis cells, antigens and energy. Let α_{ij} be specific rates of reproduction obtained at moment r for corresponding variables ($i=j$) and for reproduction by the corresponding channel ($i \neq j$); let Y_{ij} be the relative portions of corresponding clones (substances) spent on reproducing of themselves ($i=j$) and on interacting with other variables ($i \neq j$). The effect of energy factors on the immune response may be taken into account on the basis of the following assumptions: (a) the variable m_3 changes much slower than the rest of model variables; (b) the time delay τ_2 in the network "antigen - energy metabolism - immune system" is greater than the operation time of the investigated regulation mechanism; (c) there is a linear dependence of the form $m_0+m_1=M$ between the rates of appearance of cells precursors and the positive clone; (d) with the growth of m_4 the portion of energy substrates (ES) spent on the reproduction - Y_{44} diminishes; (e) with the growth of m_4 enzymatic activity in the energy reproduction - α_{44} diminishes.

Two types of problems may be solved in the framework of the model: (a) the problem of investigating the dynamics of immune response: to define the values of the functions $M_j(t)$ ($i=0,1,3,4$) over the interval $[t_0,T]$ if all model functions are specified by the prehistory $[0,t_0]$; and the rest of the functions (except for M_j) also on $[t_0,T]$; (b) the problem of minimax interaction between components of the immune system and antigen: to define the value of the functional

$$I(Y_{ij},t) = \min_{Y_{44}} \int_{t_0}^{T} \left(q_1 M_4(t) + \min_{Y_{31}} \max_{Y_{00}} \int_{t_0}^{T} [M_3(t)-q_2 M_1(t)]dt \right) dt \qquad (2)$$

where $M_i(t) = \int_{0}^{T} m_i(\tau)d\tau$, q_1, q_2 are constants,

and $Y_{00}+Y_{10}+Y_{30}+Y_{40} = 1$.

The method for the investigation of problems of type (a) for similar statements and models is given in [2].

ASSERTION 1. Let the following relations be satisfied: $\alpha_{ij}=0$ (except α_{10}, α_{11}, α_{33}, α_{34}, α_{41}, α_{44}); $\alpha_{44} Y_{44}=a_1(1-m_4)(a_2-m_4)$, $\alpha_{41} Y_{41}=a_3$, $\alpha_{34} Y_{34}Q=a_4$, $MRQ^{-1}=a_5$, $V-1+QW=0$, where $Q=(\alpha_{13}Y_{13})(\alpha_{33}Y_{33})^{-1}$, $R=(\alpha_{10}Y_{10})(\alpha_{34}Y_{34})^{-1}$, $W=(\alpha_{31}Y_{31})(\alpha_{10}Y_{10})^{-1}$, $V=(\alpha_{11}Y_{11})(\alpha_{10}Y_{10})^{-1}$. Then the modified system which takes into account the influence of energetics appears to be related to the universal deformation of the catastrophe theory - cusp, when $a_1>>\sqrt{a_3 a_4}$.

The methods for investigating problems of type (b) are given in [10]. The investigation is connected with the consecutive finding of sufficient conditions for optimality of (2) in subproblems of the type of "maximal leadership" and "stabilization" and minimization of energy consumption. For this purpose the gradient of the functional (2) is defined in the first of above-mentioned subproblems.

ASSERTION 2. The gradient $I'_{Y_{33}}(Y_{ij},t)$ of the functional (2) has the following form:

$$I'_{Y_{33}}(Y_{ij},t) = \int_{t_0}^{T} \int_{t}^{a^{-1}(t)} \alpha_{00}(t,i)Z(i)-\alpha_{33}(t,\tau) \ d\tau dt, \qquad (3)$$

where $a(t)$ is a moment of elimination of obsolete biotechnologies; $T_1 = t - a(t)$ is the averaged life span of the thymic cells; Z is the conjugate variable.

ASSERTION 3. The gradient $\dot{I}_{Y_{31}}(Y_{ij},t)$ has the following form:

$$\dot{I}_{Y_{31}}(Y_{ij},t) = \int_{t_0}^{T} \int_{t}^{a^{-1}(t)} \alpha_{31}(i,t)\,d\tau dt. \tag{4}$$

ASSERTION 4. The gradient $\dot{I}_{Y_{44}}(Y_{ij},t)$ has the following form:

$$\dot{I}_{Y_{44}}(Y_{ij},t) = \int_{t_0}^{T} \int_{t}^{a^{-1}(t)} \alpha_{44}(\tau,t)\,d\tau dt. \tag{5}$$

The general scheme of finding sufficient conditions for optimality in (3) - (5) is based on Lagrange multipliers and is considered in [10]. According to these conditions the optimal control of the corresponding relative portions in (3) - (5) will have the following form:

(a) $Y_{00}(t) = Y_{00\min}(t)$, at some part of the interval $[t_0, T]$, if $T_2 < T_1$;

(b) $Y_{00}(t) = Y_{00\min}(t)$, at some part of the interval $[t_0, T]$, if $T_2 > T_1$;

$\alpha'_{33,} \geq 0$, $\alpha'_{11,} \leq 0$, where f is a constant $T_2 = T - t_0$;

(c) at least in the neighborhood of

$$Y_{00}(t) = \begin{cases} Y_{00\min}(t), & \text{if } I_{Y'_{00}}(Y_{00},t) < 0, \\ 1 - \bar{Y}_{11}(t), & \text{if } I_{Y'_{00}}(Y_{00},t) > 0, \end{cases}$$

(d) $Y_{31}(t) = \bar{Y}_{31}(t)$ is maximally possible for $t < t_1$;

(e) $Y_{31}(t)$ is of switching character at $t > t_1$ and by turns takes the value 0 and $\bar{Y}_{31}(t)$, $M_1(t) = q_2 M_3(t)$.

In (5) the optimal control of portion Y_{44} is similar to items (d) and (e) for portion Y_{31} (15), and Y_{04} has the inverse character (Y_{04} decreases with the growth of Y_{44}).

6. Discussion of Simulation Results

The study of the cusp-model shows that bounds of the trigger domain are defined in the following manner: $m_1(a_i) < m_1 < \bar{m}_1(a_i)$, where $i = 1,2,3,5$. From the cusp-model it follows that the system is in the omneopathent domain in a state with high rates of synthesis and energy consumption, when $m_1 < \underline{m}_1$. With the growth of m_1, when the condition $m_1 > \underline{m}_1$ is satisfied the system passes into the trigger domain, remaining in the state with a high level of ES. And this will continue until the upper bound of the trigger domain $\bar{m}_1$ is reached. Then there occurs an intermittent transfer of the system into a new stationary state with low level of energy consumption. With the growth of $A = A(a_2, a_5)$ both the value of $\bar{m}_1$ and the jump amplitude are increased. Here the restoration becomes possible under more substantial limitation of the rate of energy consumption, i.e. at lower values of $\underline{m}_1$.

The optimal control on the interval $[t_0, T]$ in the immune response with the help of relative portions of cells being in the state of proliferation and differentiation is performed in the following sequence:

(a) under the action of the antigen M_3 at first $Y_{00}(t)$ is minimally admissible as corresponds to the case when $T_2 < T_1$;

(b) in the latent period and at the initial instant of the immune response $Y_{00}(t)$ is maximally close to the unit that corresponds to the case when $T_2 > T_1$, and at the next interval $[t_0, T]$ $Y_{00}(t)$ is minimally admissible; (c) the relative portion $Y_{31}(t)$ is maximally possible on the whole interval $[t_0, t_1]$ where $t < t_1$ as long as $M_1(t) = q_2 M_3(t)$, then $Y_{11}(t)$ is minimally admissible; (d) the relative portion $Y_{04}(t)$ is minimally admissible on the whole interval $[t_0, T]$ except for the section when $Y_{00}(t)$ attains its maximal possible value.

Here the case is possible when at the moment of immune system restoration (after the effect of antigen) the antigen again enters the organism. Then there arises a contradictory control strategy of the "accumulation - production" type [12]. On the one hand, for the organization of the immune response it is necessary that $Y_{00}(t)$ be close to unit; on the other hand, $Y_{00}(t)$ at the same time should be minimally admissible since it is necessary to maximize the effector function of the immune system on a small interval of time. A similar situation arises, when a stress- realizing adrenocortical hormone, hydrocortisone, enters the organism.

On the basis of the conducted study we may come to the conclusion that the strategy of iochemical adaptation is directed to the creation of metabolic conditions which strengthen the suppressor activity at initial stages of tumor growth and moderate it at later stages. The decrease of the level of ES results in the limitation of proliferation and this transfers the system from the state with a high level of suppressor and effector synthesis into the state with a low level of synthesis of the mentioned lymphocyte populations, therefore facilitating their synergy. Thus the level of ES as well as of interleukin-2 can play a part in providing the balance between individual chains of the immune system. One manifestation of the activity of this regulation mechanism is the adaptive redistribution and restorative accumulation of lymphocytes which may be observed in a thymus gland when significant doses of hydrocortisone affect the organism. It is shown (see [3]) that under the conditions of a stress which is accompanied by the change of energy status of a cell, there occurs a limitation of the processes connected with a high level of ES consumption and an activation of the processes with a low level of energy consumption. The effect of antigen and hydrocortisone reduces the adaptive possibilities, but hydrocortisone protects against great amplitudes of dropping the level of ES and m_1, thereby accelerating the restoration process (see Figure 5). By regulating the phases of differentiation and proliferation it is possible, with the help of hydrocortisone and immunocorrectors, to control the activity of the mechanism of preventing the immune system from energy exhaustion, on the one hand, and to influence the balance of individual compartments of the immune system, on the other hand.

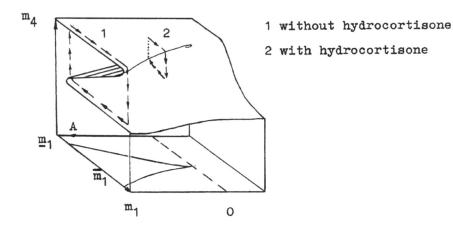

1 without hydrocortisone

2 with hydrocortisone

Figure 5

7. References

[1]   Ado, A.D. (1978) General allergology (Moscow: Medicina).

[2]   Atoev, C.L. (1984) A mathematical model of regulation of mechanical activity by creating phosphate. Kibernetika i vychisl. technika 63: 76-9.

[3]   Atoev, C.L., Ivanov, V.V. and Janenko, V.M. (1986) Interrelation of processes of synthesis and energy consumption in tissue affected by viral infection. A mathematical model. Kibernetika 3: 90-6.

[4]   Berezhnaja, N.M. and Jalkut, S.I. (1983) A biological role of immunoglobuline (Kiev: Naukova dumka).

[5]   Chum, M. et al. (1985) Enhancement of cytoxic activity of natural killer cells by Interleukin-2 and antagonism between Interleukin-2 and adenosine cyclic monophosphate. Scand. J. Immunol. 22: 375-81.

[6]   Fanta, C.H. (1985) Calcium-channel lockers in prophylaxis and treatment of asthma. Am.J.Cardiol. 55: 202B-9B.

[7]   Folkman, J. et al. (1983) Angiogenesis inhibition and tumor regression caused by heparin or a heparin fragment in the presence of cortisone. Science 221: 719-25.

[8]   Glushkov, V.M., Ivanov, V.V. and Janenko, V.M. (1983) Simulation of developing systems (Moscow: Nauka).

[9]   Gukovskaja, A.S. (1984) A role of ions in lymphocyte activation. Usp.sovr.biol. 97: 179-92.

[10] Janenko, V.M. (1986) On minimax strategy of developing systems interaction (using an organism's defence reactions as an example), in A.A. Asachenkov (Ed.) Mathematical modelling in immunology and medicine (Moscow: VINITI) pp. 148-56.

[11] Katz, A.M. (1985) Basic cellular mechanisms of action of the calcium channel blockers. Am.J.Cardiol. 55: 2B-9B.

[12] Mikhalevich, V.S., Ivanov, V.V., Janenko, V.M. and Gulling, E.V. (1986) Integrofunctional model of hemopoiesis regulation system. Kibernetika 3: 69-77,86.

Acta Applicandae Mathematicae **14** (1989), 179–189.
© 1989 *by IIASA.*

Interleukin 2 and Immune Response Control
Mathematical Model

Daniela Přikrylová

Institute of Microbiology
Czechoslovak Academy of Sciences
142 20 Prague, Czechoslovakia

AMS Subject Classification (1980): 92A07
Key words: immune response, ordinary differential equations, simulation

Mathematical modelling represents one of the methods we can use to find proper answers to questions stated to solve biological problems. This method is extraordinarily useful in cases of complex behavior of biological systems, i.e. in cases which are inconvenient to model experimentally or in cases where experimental modelling is possible but the interpretation of results requires some theoretical support. Mathematical model of the immune response regulated by interleukin 2 is presented as an example.

Features of the modelled reality

Successful immune response to a given antigen is manifested by elimination of the antigen, generation of antibody forming cells and production of antibody (if humoral response) or cytotoxic cells (if cellular immunity), and generation of memory cells. Memory cells remain in the organism after the primary response, and they ensure that the secondary response is more efficient. A characteristic feature of the immune response is a considerable proliferation of participating cells. This proliferation might be controlled by interleukin 2 which is a product of T helper cells.

Another manifestation of the immune system is immunological tolerance.

Model arrangement

This model is based on the following assumptions:
1) <u>Cells</u>:

Macrophages (Mf) remain constant, produce IL 1 after antigenic stimulation.

T helpers enter the system as Hx precursors, after antigenic stimulus they change into Ha sensitive to the second signal (IL 1, IL 2). IL 2 effects a change of Ha into the proliferating Hy, while IL 1 effects a direct change of Ha into Hz; Hy after repeated antigenic signal become Hz producing IL 2.

B cells enter the system as precursors X, after antigenic stimulation they become sensitive to the second signal (IL 2), thus changing into proliferating Y, which after repeated interaction with antigen change into Ab producing Z cells. Without meeting the antigen again, and in absence of sufficient amount of IL 2, Y cells become memory cells M.

2) Signals:

Antigen (Ag)

i) external information labeling the cell which take part in immune response;

ii) signal to the final differentiation (after proliferation and decrease of IL 2 concentration).

Interleukin 1 (IL 1) initiates the lymphocyte response via stimulation of Hx to become Hz and to produce IL 2.

Interleukin 2 (IL 2) signal for Hy and Y to proliferate.

Antibody (Ab) participate in elimination of the antigen.

The control of the immune response depends on the absolute amounts of elements engaged in it (i.e., cells, antigen, cell products) as well as on their interrelations.

Mathematical model

1) Differential equations

$$Ag' = l_a\,Ag - km_{ab}\,AbAg$$
$$(IL\ 1)' = Mf\,.\ l_i\,f_i - IL\ 1(m_i\,Ha + m_{if})$$
$$Hx' = l_x - Hx\,(f_x + l_x/Hx_0)$$
$$Ha' = Hx\,f_x - Ha\,f_a$$
$$Hy' = Ha\,f_a\,f_g + Hy\,[(l_y\,f_p - m_y) - (1 - f_p)f_y]$$
$$Hz' = Ha\,f_a\,(1 - f_g) + Hy\,f_y\,(1 - f_p) - m_z\,Hz$$
$$(IL\ 2)' = l_f\,Hz - [m_f\,(Ha + Hy + Y + M) + m_{if}\,]IL\ 2$$

$$X' = l_x - X(f_x + l_x/X_0)$$
$$Y' = Xf_x + Y[(l_yf_p - m_y) - (1 - f_p)]$$
$$Z' = (Y + M)f_y(1 - f_p) - m_zZ$$
$$M' = Y(1 - f_y)(1 - f_p) + M[(l_yf_p - m_m) - f_y(1 - f_p)]$$
$$Ab' = l_bZ - (m_b + m_{ab}Ag)Ab$$

2) Switching functions

Functions which approximate the changing probabilities of the realization of differentiation or proliferation signals were used in the standard form: $P\{Q\} = Q^2/(1 + Q^2)$. $P\{Q(t)\}dt$ is the probability that appropriate event occurs during the interval $(t, t + dt)$.

$f_x = P\{Ag/[q_x(H_x + H_a + H_y + X + Y + M)]\}$ refers to the transition from H_x to H_y or from X to Y respectively,

$f_y = P\{Ag/[q_y(H_x + H_a + H_y + X + Y + M)]\}$ refers to the transition from H_y to H_z or Y to Z respectively,

$f_g = P\{IL2/(q_gIL1)\}$ refers to the transition from H_a to H_y under condition that transition from H_a to H_y or H_z will be realized.

$f_a = P\{IL1 + IL2/[q_iH_a + q_a(H_a + H_y + Y + M)]\}$ refers to the transition from H_a to H_y or H_z,

$f_p = P\{IL2[max(0, 1 - KAg^c)]/[q_p(H_a + H_y + Y + M)]\}$ refers to the proliferating fraction of appropriate cells,

$f_s = P\{Ag/\{[q_s(H_x + H_a + X + Y + M)]\}$ refers to the production of IL 1 by macrophages.

3) Parameters

l_x...rate of precursors supply, l_y, l_a (>1)...proliferation rates, l_i, l_f, l_b...production rates, m_y, m_z, m_m...death rates, m_i, m_f, m_{ab}, k...binding rates, m_{if}, m_b, l_a (<1)...decay rates, q_x, q_y, q_g, q_a, q_i, q_p, q_s, K, c...parameters of switching functions.

Comparison of computer simulation with experimental data

As our results have shown that the consequences of the model agree with the knowledge of the immune response (Přikrylová 1985, Přikrylová et al. 1986), we started to compare the computer simulation results with concrete experimental data. From this point of view the immunological properties of two genetically different mice strains were studied. The immune response of these two strains differs in the number of generated memory cells and antibody production during the secondary response (Silver et al. 1972, Silver and Winn 1973, Říhová et al. 1981).

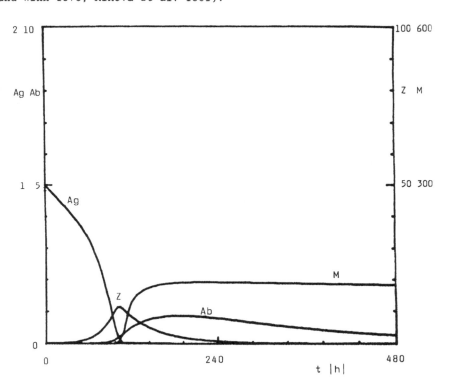

Fig. 1. High responder strain - primary response. The simulated course of the number of antibody forming cells (Z), memory cells (M), amount (in arbitrary units) of antigen (Ag) and antibody (Ab).

As a high responder strain in simulation experiments (Fig. 1, 2) we chose an example in which following parameter values were used:
$l_x = m_y = q_x = q_y = 0.001$, $l_y = 0.05$, $l_b = 0.002$, $m_z = 0.02$, $m_m = 0.00002$, $m_b = 0.005$, $l_i = 0.09$, $l_f = 0.2$, $q_a = q_g = q_a = 0.01$, $m_i = m_f = 0.07$, $m_{if} = 0.007$, $l_a = -0.002$, $m_{ab} = 0.6$, $K = 2$, $k = 0.2$, $c = 0.5$, $q_s = 0.0001$, $q_i = 0.1$, $q_p = 0.001$. Initial values: $Ag_0 = Ag_{480} = 1$,

H_x = H_{x0} = 1, X = X_0 = 1, H_a = H_y = H_z = Y = Z = M = IL1 = IL2 = Ab =0. All rate constants are related to the time scale used and have the dimension h^{-1} .

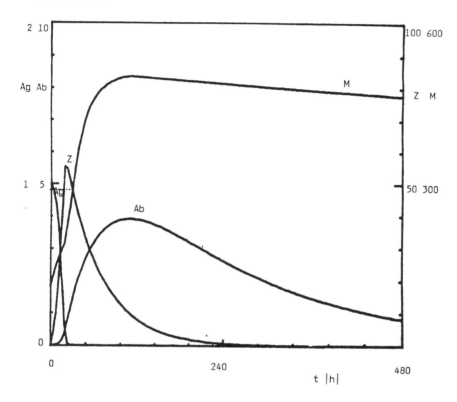

Fig. 2. High responder strain - secondary response. The simulated course of the number of antibody forming cells (Z), memory cells (M), amount (in arbitrary units) of antigen (Ag) and antibody (Ab).

As an example of low responder strain we searched for simulated immune response where the number of memory cells is lower than in the previous situation which represents a high responder strain. It is possible to obtain lower response in this sense if we assume:

1) a different character of the antigen,

2) a higher intensity of antibody production,

3) lower IL 2 production intensity,

4) lower IL 1 production intensity,

5) elongated generation time of lymphocytes

6) higher efficiency of the antigenic signal

7) different ratio of T to B precursors at the beginning of the process.

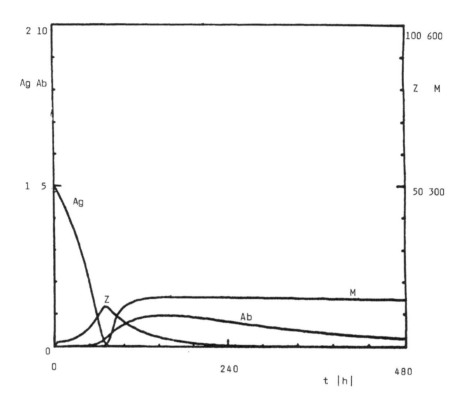

Fig.3. Low responder strain – primary response. The simulated course of the number of antibody forming cells (Z), memory cells (M), amount (in arbitrary units) of antigen (Ag) and antibody (Ab).

To choose the situation best suited to experimental data, we have compared the rate of antigen elimination – it seems that antigen elimination is faster in the case of the low responder strain (Říhová and Větvička 1984). This criterion was best fulfilled in cases 1, 2, and 7 (if the initial ratio of precursors was 5B : 1T). In the subsequent step we compared the number of antibody producing cells and antibody level generated by the system during a simulated course of

immune response. A faster nonspecific elimination of antigen leads to
the decrease of the number of antibody producing cells and antibody
level, while the assumption of a higher intensity of antibody
production leads to an higher antibody concentration but a lower
number of antibody producing cells. Only if we assumed different
initial precursor ratio, the maxima of antibody level and the number
of the antibody producing cells in primary response (Fig. 3) were
comparable with the example representing a high responder strain. This
situation (Fig. 3, 4) was chosen as a low responder strain
representation.

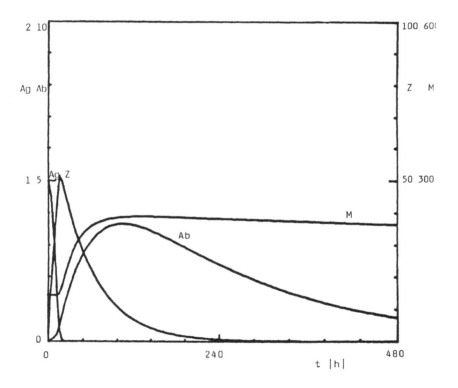

Fig.4. Low responder strain - secondary response. The simulated course
of the number of antibody forming cells (Z), memory cells (M), amount
(in arbitrary units) of antigen (Ag) and antibody (Ab).

In another example we attempted to model other experimentally obtained results - Větvička et al. (1986) have studied the effect of 5-fluorouracil on the cells of the immune system and they found that 5-fluorouracil suppressed strongly the humoral immune response. If this agent was administered concurrently with antigen or during initial 24 hours after immunization the suppression of response was up to 99 %. Moreover, it was shown that the influence of 5-fluorouracil on different mice strains differs markedly, the low responder strain being less sensitive than the high responder.

In simulation experiments we have proved our hypothesis that the different sensitivity of the studied strains to the effect of 5-fluorouracil is a consequence of a diverse course of the immune

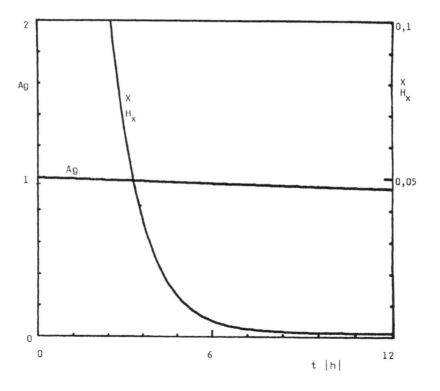

Fig. 5. High responder strain - precursors. The simulated course of the number of T helper precursors (H_x) and B precursors (X).

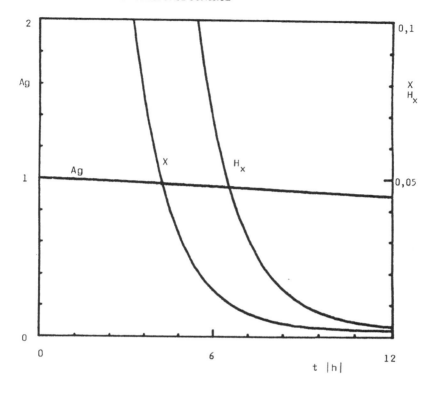

Fig. 6. Low responder strain - precursors. The simulated course of T helper (H_x) and B (X) precursors.

response during the initial hours after administration of the antigen. Figs. 5 and 6 show the decline in the number of resting cells during 12 hours after antigen administration; figs. 7, 8 illustrate the simulated course of immune response with a reduced number of precursors (we have assumed that 5-fluorouracil eliminated all cells which were activated and/or have started proliferation; after termination of the effect of 5-fluorouracil only those resting cells which were not yet activated remain in the system.

Interpretation of these results confirms the different number of precursors of cells which are involved in specific immune response as

one of the possible causes of the diverse behavior of immune systems
of genetically different mice strains.

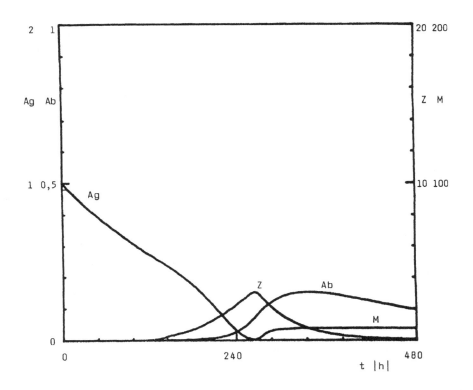

Fig. 7. High responder strain - primary response after treatment
(5 h.) of 5-fluorouracil. The simulated course of the number of
antibody forming cells (Z), memory cells (M), amount (in arbitrary
units) of antigen (Ag) and antibody (Ab).

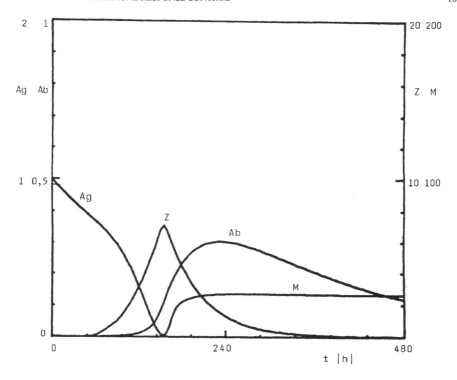

Fig. 8. Low responder strain - primary response after treatment (5 h.) of 5-fluorouracil. The simulated course of the number of antibody forming cells (Z), memory cells (M), amount (in arbitrary units) of antigen (Ag) and antibody (Ab).

References

Přikrylová, D., in Immunology and Epidemiology (Hoffman, G.W. and Hraba, T., Eds.) pp. 44-52, Springer-Verlag, Berlin 1985
Přikrylová, D., Jílek, M., Doležal, J.: Immunology Letters 13, 1986, 317-321
Říhová, B., Tučková, L., Říha, I.: Folia Biol. 27, 1981, 1-14
Říhová, B., Větvička, V.: Folia Biol. 30, 1984, 57-66
Silver, D.M., McKenzie, .F.C., Winn, H.J.: J. Exp. Med. 136, 1972, 1063-1072
Silver, D.M., Winn, H.J.: Cell. Immunol. 7, 1973, 237-245
Vetvicka, V., Kincade, P.W., Witte, P.L.: J. Immunol. 137, 1986, 2405-2410